THE BEHAVIOR-BASED SAFETY PROCESS

Managing Involvement for an Injury-Free Culture

THE BEHAVIOR-BASED SAFETY PROCESS

Managing Involvement for an Injury-Free Culture

Thomas R. Krause, Ph.D.
Co-Founder and President,
Behavioral Science Technology, Inc.

John H. Hidley, M.D.
Co-Founder and Vice-President,
Behavioral Science Technology, Inc.

Stanley J. Hodson
Technical Writer

VNR VAN NOSTRAND REINHOLD
New York

Printed in the United States of America

Van Nostrand Reinhold
115 Fifth Avenue
New York, New York 10003

Van Nostrand Reinhold International Company Limited
11 New Fetter Lane
London EC4P 4EE, England

Van Nostrand Reinhold
102 Dodds Street
South Melbourne, Victoria 3205, Australia

Nelson Canada
1120 Birchmount Road
Scarborough, Ontario M1K 5G4, Canada

16   15   14   13   12   11   10   9   8   7   6   5   4   3   2   1

Library of Congress Cataloging-in-Publication Data

Krause, Thomas R., 1944-
    The behavior-based safety process: managing involvement for an injury-free
culture / Thomas R. Krause, John H. Hidley, Stanley J. Hodson
        p.   cm.
    Includes bibliographical references (p.   ).
    ISBN 0-442-00227-0
    1. Industrial safety—Management.   2. Organizational behavior.
I. Hidley, John H.   II. Title.
HD7262.K73   1990
658.4'08—dc20   90-34384
                                                                    CIP

Contents

Preface

Growing out of a decade of consulting with industry, this book brings together several areas that we feel show great promise for the years ahead. This book represents the state of the art in the behavior-based management of safety. But more than that, it also presents a new strategy for combining training with organizational development for safety improvement. Finally, it details the implementation of a mechanism for continuous improvement which can be used in other performance areas besides safety improvement.

The 1970s and 1980s have seen the arrival and consolidation in a growing number of industries and businesses of the behavior-based approach to continuous improvement in safety performance. In addition, as the 1990s begin we see signs of wider applications of the behavior-based model presented here, to environmental issues, industrial hygiene, and quality, to name a few.

These developments represent important advances in management theory and practice. As a practical matter, safety issues have moved to a higher priority within many organizations. Furthermore, there is widespread awareness that off-the-shelf safety programs are not the answer to the needs of most organizations. Forward looking managers have learned that an effective safety effort needs to be a process, not a program, and that the process needs to be developed with site-wide representation. In this case the need for site-wide representation means that the process cannot be imposed from the top down, nor should it be understood as the preserve of safety defined narrowly as a departmental concern. In other words, the implementation of an effective safety process involves issues associated with organizational development and quality.

Behavioral methods have been used in organizational settings for many years. What is new is the strategy presented here, a strategy which draws from organizational development, behavioral science, and statistical methodology to approach a specific performance area—in this case, safety. This strategy is designed to build in methods to assure that the training that is delivered will be used on an ongoing basis so as to become part of the culture. This is a solution to the problem of transfer of learning from the training center to the workplace.

Performance related quality improvement is also sure to be part of the solution to many other management issues on the horizon. The leaner workforce, the self-directed workgroup, international competition and environmental concerns, these and related challenges call for a new strategy for continuous improvement. It is our hope that this book may contribute to the serious and informed discussion of these issues.

Acknowledgments

The authors wish to acknowledge those others, both clients and colleagues, who have contributed to the development of this book.

A consultant must admit a substantial debt to his clients. When the consulting relationship works best, both the client company and the consultant learn a great deal about the subject matter. Certain projects stand out in this regard. Monsanto Chemical Company—at both its Pensacola, Florida, and Nitro, West Virginia, locations—taught us the critical importance of employee involvement in the behavior-based safety process. For this breakthrough we are especially indebted to Paul Villane and Jim Marcombe.

Shell Chemical at Geismar, Louisiana, taught us about organizational effectiveness and the manager's role in a successful implementation effort and also first planted the idea of Trainer Training as the primary vehicle for the delivery of training content. For this important insight we are especially indebted to Pete Burright.

At Alcoa's Rockdale, Texas, facility we learned that the behavioral process could address labor management issues with a positive net result. Both Gary Griesbach and Rod Bartley played key roles in this regard.

Jim Palmer at Procter and Gamble showed us that the behavior-based process would work when implemented through a different model than our own. And Jim Spigener at Rohm and Haas Texas, Inc. in Houston, Texas, has made a significant contribution to our knowledge of organizational functioning in relation to the behavioral process.

BST consultants have also contributed to this book, especially Dr. Kim C. M. Sloat who has successfully managed implementation of the behavior-based safety process with consistent high quality, distinction, and new ideas for improvement. Don Groover helped draft the material on accident investigation.

THE BEHAVIOR-BASED SAFETY PROCESS

SAFETY PROCESS

Managing Involvement for an
Injury-Free Culture

PART 1. INTRODUCTION

Chapter 1

The Development of Behavioral Technology as a Tool for Continuous Improvement in Safety Performance

-

As often happens with basic scientific research, a number of years went by after the discovery of the principles of behavioral science before they were applied in industry. B. F. Skinner did his pioneering work in behavioral science in the 1940s, but it was not until the late 1960s that these principles found wide application in institutional and educational settings. In the early 1970s Skinnerian behavioral science found various industrial applications. The behavioral approach is valuable in numerous areas of industry, and among them is safety. The safety application is the subject of this book, and it is the field that the authors have worked in exclusively since 1980.

The major industrial application of scientific methodology is in the field of quality—Deming's work provided the impetus here. The widespread application of scientific methods to industry is probably an idea whose time has come. Certainly the use of statistical process control (SPC) has shown managers across industries tremendous potential benefits across the board — not the least of which is a change in management style itself. And given the success of behavior-based safety management, the application of behavioral principles in other areas of operations and human resources is only a matter of time.

Behavioral science as we know it begins with the work of the Harvard psychologist, B. F. Skinner. Before Skinner's work other psychologists had paid attention to behavior, not as a subject in itself but as an indicator of *internal states.* This was because most researchers felt that internal mental states—attitudes, states of mind, perceptions, etc.—were the most important subject for psychology. Skinner, however, was very interested in what could be known about behavior as a subject in its own right, and in the 1940s he pioneered the rigorous application of the scientific method to the study of behavior. Skinner used strict measurement to identify and vary the elements of animal behavior under controlled conditions. In this way he produced the first systematic, verifiable findings about overt, observable behavior applicable to people as well.

Skinner's basic approach represented a radical departure from the way psychology was viewed at the time. Skinner developed techniques for measuring behavior, and using these measures he was able to demonstrate scientifically valid laws defining the relationships between behavior and other factors. Primary among these factors was the role of *consequences,* events which *follow* behavior. Skinner's discovery of the primary significance of consequences came as a surprise both to our intuition and to the science of the day. This discovery showed that *external* factors could reliably predict behavior. This was one of Skinner's most important additions to our store of knowledge about ourselves. Before Skinner's discoveries, most people took it for granted that internal states were the most important factors determining how we behave. Having established that external factors such as consequences had the ability to control many behaviors, Skinner further defined consequences. He identified a type of consequence which he called a *reinforcer*—an event which not only follows a behavior but which increases the probability that similar behaviors will occur again in the future.

Over the years Skinner did a great deal of work in this area, publishing many scholarly articles and establishing the fact that scientific methodology could be applied to the study of human behavior. Out of this work grew the general field of study which is known as *applied behavioral analysis.* Since the time of Skinner's early work this has become a flourishing field of research and scholarly publication. In addition, applied behavioral analysis has given rise to the application known as *behavior modification.* Behavior modification represents this basic work directed to human beings, particularly those in mental health and educational settings. Once again, a sizable scientific literature has been developed in behavior modification. The literature on behavior modification documents the application of these basic principles in a wide range of clinical and performance areas, including their role in the treatment of alcoholism and drug addiction, various kinds of neurotic and psychotic disorders in both adults and children, and virtually every other type of mental disorder. Generally speaking this approach has been effective and has the advantage of having demonstrable scientific validity. It is most useful when the disorder is primarily characterized by its behavioral component.

The early work with behavior modification in industrial settings has shown excellent potential based on results, but it has never really gained popularity and in some cases behavior modification has a poor reputation. This is due primarily to the inherent difficulty of proper implementation. While the basic ideas are straightforward and sound, their implementation is beset with problems. When behavior modification is presented incorrectly, it can be perceived as manipulative. Or where employee involvement is not adequately engaged, the behavioral approach is misunderstood as just another program. Similarly, if management fails to understand the principles of the behavior-based safety process, they unknowingly subvert the Implementation effort. On the other hand, as the quality improvement process gained support and visibility beginning in the 1980s, the central role of behavior in performance has been highlighted, causing a renewal of interest in the behavioral approach.

As the field of behavior modification matured, it was applied in industrial settings. This work has been well summarized by F. Luthans and R. Kreitner (1975). Before 1978, the majority of the work of behavioral analysis in industrial settings was related to various aspects of productivity. However, 1978 was the year that marked the systematic application of these methods to the study of industrial safety. In that year Dr. Judith Komaki, then at the Georgia Institute of Technology, published the first serious study using applied behavioral analysis for industrial safety. Dr. Komaki's interest was not solely in safety; she was interested in the use of applied behavioral analysis techniques in industry in general. However, safety became the focus of her study because of one of her students. As part of his course work, one of her students proposed to use behavioral techniques to design a performance improvement process at his family's 200-employee food processing plant (personal communication).

As it happened, safety was the area in which the plant most wanted to see performance improvement. Before Komaki's study there had been other studies directed at safety using various aspects of behavior modification, but not one of these studies was sufficiently thorough. Komaki's study was unique in that it involved the systematic application of applied behavioral analysis to the subject of safety in the work place. Her study was comprehensive in that it:

- Involved the definition of generic safety-related behaviors—identified as most prominent in the work environment.
- Defined the safety-related behaviors in operational terms
- Based the Observation of these behaviors on preset operational criteria
- Administered feedback to the workers on the basis of the variability of observed safety-related behaviors.

Before Komaki's study set the stage properly, there had been partial applications of behavioral technology to industrial safety, but they were seriously flawed. It is an ironic fact that a method of safety sampling had been developed and used in a number of companies years earlier. However, the data generated from safety sampling was generally based on a poor measurement system, and the data was maintained confidentially by the safety office! A typical failing of these early sampling techniques was that they produced only raw scores of safe or unsafe behavior with no measure of the *rate* of occurrence or the *proportion* of safe to unsafe behavior. Very similar errors are found in the early attempts to apply behavioral techniques to accident prevention. In contrast, Komaki's approach was methodologically sound, and the result was a method for the measurement of the proportion of safe behavior present in the work place. This measure of proportion is crucial to satisfactory performance. If the measure taken is of either safe or unsafe behavior alone, then the measure is necessarily confounded with the overall activity level of the workers observed. Following Komaki, numerous studies have appeared using this basic approach. (See the bibliography for references.)

The authors' work in this field has been in the application of behavioral methods for accident prevention in industrial settings, not for research purposes but with the emphasis on implementation issues. The objective of this work is continuous safety performance improvement, organizational development, and training in critical skills.

THE IMPORTANCE OF THE ORGANIZATIONAL DEVELOPMENT APPROACH

Over the years most training in behavioral methods for industrial applications has taken a generic approach. Instead of training entire work units, the standard procedure was to select managers and supervisors, train them in the use of behavioral methods, and then hope that this training would be applied successfully in areas left to be chosen by the manager or supervisor himself. The problem with this approach is that although the managers and supervisors are not given any training in implementation, they are nonetheless expected to *transfer* their generic lessons to specific areas of performance. The result is a predictable loss of translation between the concepts taught— usually by academics—and the eventual real-world applications. The other failing of this approach is that the trainee, left to his own resources, is strongly tempted to adopt a program from somewhere else rather than adapting what he has learned.

This is a common problem with industrial training of all kinds. This training gap accounts for the fact that many supervisors and managers may have had repeated management training courses in the same material and still do not know how to apply the principles of that material. To make matters worse yet, the management system within which the managers are indeed trying to apply their behavioral knowledge may not support them; it may even undermine their efforts at application.

For these reasons the safety management process presented in this book has many similarities with the organizational development model, an approach which addresses training and implementation issues at the same time. This approach involves the initial evaluation of a performance area—safety improvement in this case—and the subsequent development of training and organizational methods to support the achievement of performance goals. In this way a *structure* or system or process is created which senior management can support and in which employees can be trained to take their respective roles and responsibilities. The structure is the behavioral process—identify behavior→measure performance→give feedback→identify new behavior, and roles are defined based on employee level within the organization. Training is provided to enable the employees to perform their roles in the overall structure.

In this approach, the role of the outside consultant is to provide training— usually through Train-the-Trainer course work and hands-on workshops— and consultation on the phases of the implementation effort. In this way a bridge is built to cover the gap described above. The consultant is working with the organization that is actively *engaged* in implementation, not merely telling the organization *about* implementation. The effort at organizational development which the continuous improvement safety process requires of most facilities is one of the valuable "side effects" of implementation. The facilities that take to behavior-based safety management most readily are the ones that are already structured to use the SPC approach to quality improvement. This is not surprising, considering the many similarities between these two process approaches to safety and quality respectively.

PARALLELS WITH THE QUALITY IMPROVEMENT PROCESS

Although the behavior-based approach to safety management is relatively new, many of the managers that the authors have worked with say that it reminds them of the quality improvement process. This is an accurate observation. The quality improvement process and the behavior-based accident prevention process have many similarities, and their most important similarity is that both of these intervention systems derive from basic scientific methodology. While Deming and other industrial statisticians have applied scientific methodology to quality, the authors have applied scientific methodology to safety-related behavior in the workplace. Common to both of these approaches one finds:

- Operational definitions
- Measurement
- The display of data
- Feedback
- The use of data to quantify, measure, and provide information enabling workers to determine relationships between variables
- Problem solving
- The role of the manager is to initiate and support process improvements

These techniques represent the single most efficient methodology for safety performance improvement in the workplace.

IMPLEMENTATION IS THE KEY

While the principles involved are straightforward, both in safety and in quality, the application of these principles is anything but straightforward. In fact, in order for an organization to apply these principles in the long term, the organization must either be functioning at a high level of effectiveness or else be willing to address the obstacles to high-level functioning. It takes an efficiently structured, highly effective organization to be able to:

- Involve its employees in the commitment to improve safety
- Develop credibility regarding this safety commitment
- Draw up operational definitions of critical behaviors
- Teach its employees to gather data on these critical behaviors
- Display, understand, and respond to that data systematically so as to identify and solve problems

Effective organizational functioning means a number of important things. Communication is one of them. The workers must be able to communicate with supervisors, and departments need to communicate and cooperate with each other. The workers must be able to see that overall, group performance improvement benefits them individually and thus deserves their cooperation in the group effort. In general, group processes need to take place with efficiency, and there must be trust and credibility between management and workers. The importance of these factors to the success of the performance improvement process explains why Deming, for instance, has not limited his work to statistics but has also developed management training ideas and a management philosophy about how to use statistical principles effectively. Deming has been drawn into this further work because a wide variety of management issues must be addressed for the proper implementation of a scientific approach to quality improvement. The authors have found that the same thing holds true for the implementation of the behavior-based accident prevention process. In this sense, the implementation of behavior-based safety management is a somewhat unique combination of training and organizational development. Organizational development is involved in the sense that the organization looks systematically at how it functions and then designs intervention strategies for improvement—creating better communications by breaking down barriers, encouraging more openness and trust, driving out fear, building teamwork, etc. All of these changes are changes for the better and so it is no surprise that many companies who have committed themselves to these new approaches in safety have found that the "side effect" of cultural change has been as helpful to them as the benefits in terms of accident prevention.

A case in point is the Procter and Gamble Company, probably the first U.S. company to employ behavioral strategies for safety in a substantial way. Setting out on their own, they began to develop these techniques in the mid-1970s and have further developed and perfected them over the years. They still use them throughout their organization. Compared to Komaki's approach, which was an informed elaboration of known principles, Procter and Gamble developed their own approach based on basic behavioral principles. Their results have been outstanding and clearly they have demonstrated the long-term effectiveness of the behavioral process. And in keeping with the affinities between safety and quality management, in recent years Procter and Gamble has moved its safety behavioral strategy toward quality as well.

SUMMARY

Since the late 1970s, numerous major facilities have taken on the behavior-based approach presented in this book, using it to achieve long-term results. Starting in 1984, Monsanto has been a pioneer in the use of behavioral methods for accident prevention. Monsanto's very successful Implementation effort has shown that employee involvement and participation is central to success.

Shell Chemical Company has also done pioneering work in the area, with long-term success. In the late 1980s, many companies were in the first or second year of development of the behavioral process. These included Alcoa, Rohm and Haas, ARCO Chemical Company, The Pillsbury Company, Scott Paper Co., Georgia Gulf Corporation, Chevron U.S.A., and Boise Cascade Southern Operations. At the time that these companies Implemented the behavior-based process, they all had excellent existing safety performance. They also had a vision of an injury-free environment and an awareness that behavioral issues had to be dealt with adequately if they were to achieve their vision. They saw how the foundation concepts of the behavioral approach could be adapted to their facilities through the Assessment and the Implementation efforts, presented in Part 2 and Part 3, respectively of this book.

The next chapter presents the core concepts of the behavior-based accident prevention process.

Chapter 2

Foundation Concepts of Behavior-Based Safety Management

The behavioral approach to safety management applies principles of behavioral science in several important areas. During the Assessment phase, this approach uses behavioral methods (Interview techniques, Antecedent-Behavior-Consequence (ABC) Analysis, surveys) to analyze the strengths and weaknesses of existing safety measures and of safety culture and to target the most important opportunities for improvement on the basis of a facility's incident reports and records. During the Implementation phase, behavioral principles of work pattern-search and operational definitions are used to develop an inventory of behaviors that are critical to safety at the facility, and the sampling and charting of the facility's performance of this set of critical behaviors establishes a baseline measure of the existing levels of percent safe behavior within the sampling area. This concept of sampling the *mass* of behaviors in the target area is central to the behavioral approach, and it is presented in greater detail below. Finally, after the workforce has been introduced to the baseline figures for its performance of the critical safety-related behaviors, verbal and charted feedback are used to modify and improve percent safe performance of the facility's mass of safety-related behaviors.

In the same way that Deming and others have applied scientific research methodology to improve quality (operational definitions, measurement, feedback), the behavioral approach to performance applies the same methodology to a variety of behavioral issues—in this case the issue of safety performance. This similarity of origin accounts for the strong parallel between behavior-based safety management and the quality improvement process. For more on the parallels between safety and quality, see *Measurement* in this section of the book.

This chapter ranges over some of the basic concepts and practices of the behavior-based continuous improvement process, addressing first what it means to take a critical-mass approach to unsafe behavior. The discussion then turns to the important topics of how to improve safety culture by focusing on behavior versus attitude and how to ensure employee involvement in the safety process. The closing presentation sketches Antecedent-Behavior-Consequence Analysis, a powerful management tool for discovering the existing strengths and weaknesses in a facility's safety efforts.

THE CRITICAL-MASS APPROACH TO UNSAFE BEHAVIOR

At the outset, the most important thing to understand about the behavior-based approach is that this approach focuses on the sheer *mass* of unsafe behaviors at a facility. The unsafe behaviors in question are the work practices of the facility. In a majority of cases—from 80% to 95%—accidents are caused by unsafe behavior. This statement emphatically does *not* mean that the injury is the employee's fault. Nor does this statement contradict the diagnosis of quality improvement personnel that 80% of the problems with quality are due to poor management practices. Wherever it appears in this book, the statement that 80% to 95% of accidents are *caused* by unsafe behavior refers to a very specific kind of cause known as the *final common pathway*.

For example, a worker may be feeling pressured by the production schedule, and at the same time perhaps he or she is preoccupied with a daughter's illness. However, if the worker gets hurt during this time it is almost always because the worker *does something unsafe* in response to the situation—some risky action such as trying to clear jammed equipment without first turning it off, for instance. In other words, the worker's anxiety about production pressures and family worries are understandable facts of daily life; however, the concern here is not with them but with the Observable, risk-taking behavior of reaching into moving equipment. This behavior is a *critical behavior*—so-called because in this case it is a behavior that makes a critical difference in whether or not a worker gets injured while using the equipment in question.

The statistics of the work environment make it clear that a very large number of unsafe acts and/or conditions precede every accident. To use an image, this swarm of pre-existing unsafe behaviors and conditions is in the air, like water vapor just waiting to precipitate out as a thunder shower or like the avalanche primed to happen. Focusing on behavior is crucial because an instance of unsafe worker behavior is like the small sound that touches off the avalanche. These small causes that precipitate large effects are causes that provide a final common pathway for many preceding causes to come together. This is the type of cause that unsafe employee behavior represents. In 80% to 95% of all accidents, employee behavior provides the last link and common pathway for an accident to happen. However, the unsafe behavior at issue is a part of the management system, implicitly either encouraged or condoned by management. Therefore, to blame employees is counter-productive.

The effective approach is a behavior-based process approach that identifies critical safety-related behaviors, measures the sheer mass of them, and manages their levels so that a workforce stops precipitating accidents. This concept is new in safety circles, but it has gained a recent foothold in quality improvement efforts, and it has been known and used in science for a century now. In order to picture the linkage between unsafe behavior and injuries, it is helpful to think of the relationship between the critical mass of a radioactive substance and its explosiveness. For instance, in the case of a piece of uranium of critical mass, no one has any idea precisely which unstable atom is going to touch off the chain reaction resulting in an explosion. Just as there is a randomness or unpredictability about individual atoms in a mass of uranium-238, so there is a randomness and unpredictability about individual employee behaviors at a particular facility. On the other hand, the activity of the whole mass of uranium is *statistically* very predictable. And in the same way, the overall safety performance of an entire facility is statistically very predictable; at a given level of unsafe behavior there will be explosive events — accidents are going to follow.

Given what physical science knows about radioactive uranium, the people who manage it are very careful to store it in quantities well below its critical mass threshold. Given what behavioral science knows about behavior, managers responsible for safety do something similar. They identify the accident threshold (the critical mass) of their facilities by conducting a behavioral analysis to identify critical behaviors. Once they have identified the critical behaviors, they measure compliance through Observation and they provide feedback for improvement. This core linkage — facility safe behavior percentage up, facility accident frequency down — has been demonstrated in a variety of business and industrial applications. The first two graphs (see Figure 2-1 and Figure 2-2) from a nylon producing facility, illustrate the inverse relation between %Safe levels and accident frequency. These results are typical of behavior-based safety management efforts. For case histories, see Chapter 17, and for an overview of industry-applied behavioral research, see Chapter 1.

In each facility the rise and fall in the level of safe behavior is a function of various factors: management system, workforce, physical plant, machinery and processes of production, the product itself, etc. As the frequency of unsafe behaviors increases, the likelihood that injuries will occur increases. Many first aid injuries generally occur prior to a more serious injury, and so forth up to fatalities, as shown in Figure 2-3.

The challenge is to determine the accident thresholds of a particular facility and then to track and manage worker behavior at levels well below even the first threshold. This proactive management of safety performance steers by indicators in advance of even first aid accidents. To steer by injury levels even as relatively benign as first aid accidents is to give up management control and to invite the fluctuations of the accident cycle. For a presentation of the accident cycle, see Chapter 3. The accident cycle is a familiar fact

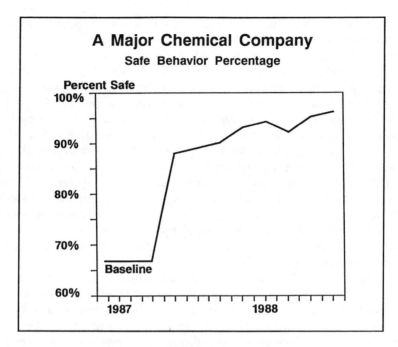

Figure 2-1.

of organizational life where an organization's safety effort is based on response to injuries. This approach is reactive rather than proactive, and it produces an intermittent, marginal solution to the problem. By waiting for recordable accidents or injuries to trigger its response, the reactive approach proceeds as though continuous improvement were impossible.

Whether hidden or explicit, the premise of reactive safety management is that staying even is all that is possible. Behavior-based safety management focuses on the strategic, long-term needs of a company by perceiving safety as a long-term product of the organizational system rather than as an accidental by-product of unknown origin. It is also important to distinguish the behavioral approach from other approaches which emphasize attitude change over behavioral change. As its name implies, the behavior-based approach improves safety culture by identifying and then managing a change in the behaviors which are critical to safety in a given facility.

BEHAVIOR VERSUS ATTITUDE

People are more inclined to pay attention to change in the opposite direction, the attitude change that brings a change of behavior. It is because attitude has the power to affect behavior that people are tempted to focus on it in the

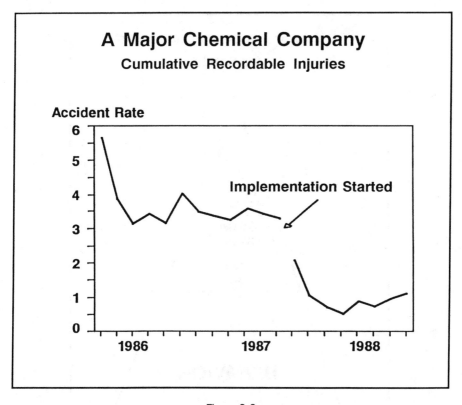

Figure 2-2.

first place. If a change in safety-related behavior is desired, slogans are posted and meetings are held urging people to change their attitude about safety. Behavioral science points out that if managers want to change behavior, there is good news—it can and should be done, *starting* with behavior. It turns out that there is a two-way street between attitude and behavior, and behavior has the power to change attitude, too. And in a business or industrial setting there are two powerful reasons to focus on behavior first:

1. Behavior can be measured and therefore managed, whereas attitude presents measurement problems.
2. A change in behavior leads to a change in attitude.

Although programs emphasizing attitude change make a strong appeal to our common sense, such programs are flawed because they overlook these two points. Everyone agrees that a good safety attitude is important. Given the importance of measurement in the management process, it follows that

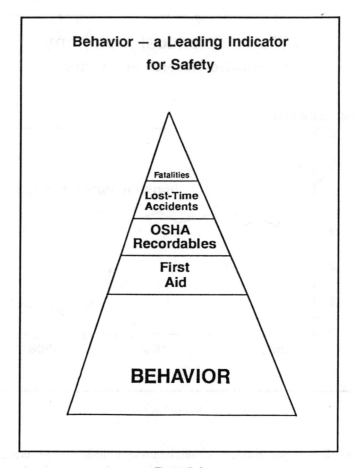

Figure 2-3.

attitude change is very difficult to manage. To actually manage a process of continuous safety performance improvement solely by means of attitude change would require some form of attitude monitoring and control, which is not feasible even if it were desirable. In actual practice, attempts at changing safety culture by changing attitude invariably suffer from a lack of control.

The key here is that culture cannot be *systematically* changed by focusing on attitude change. For instance, to the question, *What are we doing to improve safety attitude, and how do we know whether it works?* a prudent manager would be hard pressed to say anything more than, *We try many things—it's hard to tell what works and what doesn't.* But to the question, *Is the incidence of unsafe body placement on the increase or the decrease at our plant?* a behavior-based safety management process can give a definite answer, indicating by how many percentage points the incidence of the specified behavior is up or down over the last measurement period. The objectively observable character of behavior makes it amenable to both measurement and management, addressing the first weakness of attitude-focused programs. The second strength of behavior-based management is that changed behavior causes a change in the attitude of the workforce.

Although people are not usually aware that changed behavior can change attitude, this effect of behavior on attitude is also a common fact of life. In domestic life parents make use of the behavior-to-attitude directed linkage whenever they accustom their children to a certain household routine — *Clean your bedroom on Saturdays, and remember that Sunday is a school night.* Regardless of what the children may think (attitude) about these routines at first, after a while they begin to feel uncomfortable (their attitude changes) if they do not get a chance to restore the order of their room (behavior) on Saturday or to look at their homework on Sunday night before the week starts. This is called the development of good habits, a process that works because a change in behavior can make a difference in attitude.

Where adults are concerned, a recent example of this behavior-to-attitude change in American life has to do with the use of seat belts in automobiles. As recently as the 1960s, seat belts were something of a rarity in cars. When seat belts were first introduced into cars, many people had driven for years without them. In the early days, drivers polled by researchers about seat belt use said that they did not use (behavior) seat belts because they felt uncomfortable (attitude) wearing them, and they worried (attitude) that the belt would somehow trap them in their cars in the event of an accident. Many drivers had a negative attitude or feeling about seat belts, and this was reflected in their behavior — they did not voluntarily install seat belts in their cars, nor did they use seat belts even when they were already installed in their cars. Over the years seat belt use has changed from being a novelty, to being a recommended option, to being required by many businesses, municipalities, and state highway patrols. People have been encouraged to change their habits (behavior) no matter what they might think (attitude) about seat belts. After only a short time of regular seat belt use, research showed that many of the same drivers polled earlier said they now felt uncomfortable when they *did not* use seat belts. Their attitudes about seat belt safety had changed around completely. This is a classic instance of the way that a change in behavior can cause a change in attitude.

This same principle has been demonstrated in the work place on numerous occasions. For instance, during the Assessment effort at one facility, behavioral analysis of the accident reports revealed that one of the behaviors associated with a significant number of accidents and/or injuries was *unsafe body placement in relation to task.* Here was a prime candidate for the facility's inventory of critical behaviors, one that can be targeted as a leading indicator of how safe or unsafe the work place is, in advance of any accidents at all. The rationale is straightforward—the more times that workers use unsafe body placement in relation to task, such as standing in the line of fire, the more likely it is that an accident is going to happen. Conversely, other factors being equal, the fewer times that somebody stands in the line of fire, the less likely it is that an accident is going to happen. Using a Data Sheet that incorporates the operational definitions of such critical behaviors, trained Observers make random samples of work place behavior, producing a measure of the facility's level of safety performance.

At the facility in question a common sequence of events took place. Work crews typically take their unsafe behavior for granted. Within the safety culture of this facility there was the unspoken assumption that unsafe body placement was simply part of being an efficient worker, a team player with hustle. Within six to eight months of experience with the behavior-based safety process, however, their behavior and their attitude changed dramatically. The workers came to understand the basic concepts of the process. They saw that the Observers were careful to keep their observations objective and accurate, they saw that management was careful not to distort the observations with disciplinary action or punitive measures. The workers saw that their performance was being tracked against standards that *they and their peers* helped to define during the development of the facility's inventory of critical behaviors. This is the constant message of the behavior-based process. The result is that the same workers who used to think of standing in the line of fire as an acceptable thing to do come to regard it, and related unsafe body placement, as irresponsible behavior that is improper in a mature team player.

ENSURING EMPLOYEE INVOLVEMENT

The behavior-based approach to accident prevention builds in this kind of workforce involvement. From the outset, site-wide input is used to draw up and define the inventory of critical behaviors for the facility. Properly administered, this process means that crews become interested in their safety performance curves relative to their own past ratings and relative to the ratings of other crews. Where the supervisors may have felt in the past that they were always having to nag their workers about safety, they now find that the workers initiate discussions about their safety performance ratings. When they achieve several periods of unbroken improvements, they workers are proud of themselves. Then when the graph of their safety performance shows a decrease in the percentage of safe behaviors for their crew, they want to know exactly which critical behaviors are responsible for the dip in their good record.

The "secret" of this kind of participation and involvement is that behavior-based safety management avoids personalities by focusing on something that is objectively measurable. Although it seems a bit paradoxical at first, it is because the behavioral approach appreciates attitude change that it does not waste time and resources ordering people to change their attitudes. This is a fruitless exercise because people cannot obey such orders, even if they want to. It is a fact of human makeup that people cannot be commanded in attitudinal matters. People cannot be commanded to change their attitudes any more than they can be commanded to hope or to believe or to like something. They can, however, manage not to stand in the line of fire while they are working. Therefore, the first step in developing this safety process is to make an accurate analysis of the safety-related behaviors that are actually occurring in an organization.

ANTECEDENT-BEHAVIOR-CONSEQUENCE (ABC) ANALYSIS

An essential tool of safety management is discovering and addressing the roots of accidents. It is well known that 80% to 95% of accidents are caused by unsafe behavior. All safety efforts that work—whether or not they are consciously behavioral in their orientation—are effective because they influence employee behavior. Equally crucial is that most organizations have what amount to very strong behavioral incentives that favor *unsafe* behaviors. Behavioral analysis helps the organization to assess the factors that are really driving its safety efforts. This basic tool of behavioral analysis is known as ABC Analysis, and it provides the powerful foundation of behavior-change technology. In terms of this analysis, an *antecedent* is an event which triggers or elicits an objectively observable *behavior.* A *consequence* is any event that follows from that behavior. An example is the ringing doorbell (antecedent) which we answer (behavior) so as to see (consequence) who is at the door. Common sense tends to identify the antecedent, in this case the doorbell, as the most powerful stimulus to behavior, in this case answering the door. And of course the antecedent is important. However, in his pioneering work in behavioral science, B. F. Skinner showed that consequences are more powerful determinants of behavior than are antecedents.

To see the truth of Skinner's discovery, suppose a situation where the doorbell rings repeatedly and repeatedly there is no one at the door. Perhaps the bell is malfunctioning, or pranksters are ringing the bell and then running away from the door. In such a case, the behavior of answering the door to see who is there is frustrated for lack of the expected consequence. In fairly short order one would stop "automatically" answering the door. As soon as the ringing doorbell no longer reliably signaled the presence of a caller at the door, it would no longer elicit in us the behavior of going to the door to see who was there. Taken by itself, the antecedent (the bell) does not directly determine the behavior (answering the door). Instead antecedents elicit certain behaviors because they signal or predict consequences. It is the goal of ABC Analysis to discover which consequences are controlling a particular behavior. Once the controlling consequences are known, they can be changed; and when the controls change, the behavior changes. Where safety performance is concerned this means finding out which consequences are actually driving behavior in the work place.

In a nutshell, ABC Analysis involves the following principles:

- both antecedents and consequences control behavior,
- but they do so very differently,
- consequences control behavior powerfully and *directly,* and
- antecedents control behavior *indirectly,* primarily serving to *predict* consequences.

Many well-intended safety programs fail because they rely too much on antecedents—things that come before behavior—safety rules, procedures, meetings, etc. All too often these same antecedents have no powerful consequences backing them up.

In the ABC Analysis of real-world situations, one finds that most behaviors have a cluster of consequences which follow from them. For instance, the following is a brief but representative list of consequences commonly cited by workers regarding failure to wear hearing protection:

- greater comfort when they are not wearing the protective equipment,
- greater convenience in not having to locate the protective equipment and put it on, and
- the possibility of hearing impairment.

Each of the natural consequences in this set of consequences is like a plus or a minus, competing among themselves to determine what behavior the worker will exhibit the next time he does a job that requires hearing protection. The antecedents in the environment—the sign that reminds the workers to wear hearing protection, the supervisor who mentions it now and then, the high-pitched or high-intensity sound of the machinery itself—these antecedents have a less direct influence over the workers' behavior than the consequences which follow from their behavior. With this in mind it is important to note that the first two consequences listed push the worker toward *not* wearing protective equipment for hearing, while the consequence of a hearing injury pushes him toward wearing it. In addition to discovering that consequences are stronger than antecedents, behavioral science research has also found that in the competition of consequences to control behavior, some consequences are stronger than others.

Soon-Certain-Positive—The Strongest Consequence

There are three features that determine which consequences are stronger than others:

Timing. A consequence that follows soon after a behavior controls behavior more effectively than a consequence that occurs later.

Consistency. A consequence that is certain to follow a behavior controls behavior more powerfully than an unpredictable or uncertain consequence.

Significance. A positive consequence controls behavior more powerfully than a negative consequence.

These three rules mean that the consequences which have the most power to influence behavior are consequences which are simultaneously soon, certain, and positive. By contrast, the weakest consequences are the ones that are late, uncertain, and negative. Because they are so common, and often so ordinary, examples of the first kind of consequence can seem trivial and not worth paying attention to. That is the point, however; such consequences are so common because they are so powerful in spite of their seeming triviality. The example about protective equipment for hearing is a case in point. A safety program that tries to motivate the use of personal protective equipment solely by stressing the possibility of hearing loss is relying on the weakest kind of consequence—one which occurs slowly or eventually (later), if it happens at all (uncertain), and which is negative. Common sense may think that it is illogical to risk something as serious as hearing loss for some very small but immediate convenience or comfort, such as not having to locate and wear protective equipment for hearing. The fact is that in the give and take of everyday situations most people continually take such risks for precisely these small but immediate, certain, positive outcomes. Whether it is common sense or not, it is human nature to behave this way. From a long-range point of view, it is clear that a safety effort that relies solely on the possibility of hearing loss to influence workers to use protective equipment for hearing is a safety effort that is going to sustain hearing loss.

Consider two managers, one who wears hearing protection when it is required, and one who does not. What difference between these two is most likely to explain the difference in their safety-related behavior? Not their nature, and not the antecedents. They have the same human nature, and they are both aware of the antecedents of their behavior—that is, they both *know* that they *should* wear protective equipment when it is required. One manager wears protective equipment for hearing and the other does not and may not even think about it. The difference in their behavior is most likely to result from different sets of consequences for their behavior. The manager who is careful to wear hearing protection very probably receives strong reinforcement for behaving in an exemplary fashion. This manager is part of a safety culture that defines this as good performance and consistently gives rewards for it. The safety culture of the other manager is very likely one that offers no support or rewards for paying attention to hearing protection in particular and for guiding safety-related behavior in general.

The two managers have different attitudes about safety because their attitudes internalize and reflect their respective safety cultures, and their safety cultures differ in what they reward and what they punish. The first manager feels the importance of safety and experiences a sense of leadership when setting a good example. This manager's attitudes about safety have become an internal source of soon-certain-positive reinforcement for safety-related behaviors. The difference between these two managers highlights the importance of the safety culture as the primary source of consequences that build the attitudes which in turn guide and reinforce safety behavior.

It is not the point of behavior-based safety management to change human nature, but rather to change the safety culture, to use the nature of behavior in *favor* of safety instead of against it. This amounts to devising consequences for safety that are soon, certain, and positive. This is the Implementation effort, and it is a bit like priming a pump. The initial soon-certain-positive consequences in favor of safe behavior build new attitudes toward safety. These new attitudes in turn become the source of both broader and more finely tuned attention to safety, bringing new soon-certain-positive consequences to bear on the facility's safety performance. In this way the positive and continuous improvement process is established in a facility, becoming a safety mechanism.

Subsequent chapters of this book detail the ways of implementing these powerful measures in favor of safety, primarily in the form of verbal and charted feedback for the workforce on the basis of improved safety performance.

The range of possible consequences.   Between the most powerful consequences (soon-certain-positive) and the weakest ones (late-uncertain-negative) there is a range of consequences that are mixtures of these three characteristics. For instance, among the consequences of not wearing protective equipment, comfort and convenience were of the most powerful kind, whereas the possibility of hearing loss was of the weakest kind. This is the most clear-cut case. Oftentimes, however, each of the consequences of a particular behavior is a mixture of strong and weak effects. The following three examples of cigarette smoking, insurance premiums, and traffic tickets are presented both for their familiarity and because they have near relatives in the work place.

In the case of habitual cigarette smoking, for instance, on the one hand the consequence of lung impairment is powerful because of its near certainty; but on the other hand it is negative (weak), and for most people it comes years after (late/weak) they have begun to smoke. In the terms of ABC Analysis this consequence gets a middling rating because it is one that is not-soon (weak), certain (strong), and negative (weak). No wonder then that for many people the immediate, certain, positive satisfaction of physical craving they experience in smoking outweighs the prospect of respiratory troubles. Given the laws of human behavior, this seemingly small pleasure is a very powerful consequence in favor of smoking.

Insurance policies present another case of a consequence that is a mixture of strong and weak effects. Most, if not all, insurance coverage offers a consequence which is clearly positive (strong), but much of this coverage is something that the policyholder may never need (uncertain/weak), and which may come years after a person begins paying premiums (late/weak). In the face of these odds, it can be much more pleasant (positive/strong) to spend the premium money on some known good (certain/strong) now (soon/strong).

When drivers receive traffic citations for speeding, they experience a consequence that happens as they are apprehended in the act (soon/strong). Receiving the ticket, however, is a negative consequence (weak), and the odds of receiving one are uncertain (weak) given the relative scarcity of highway patrol officers. Therefore it is no surprise that many drivers exceed the speed limit, since by doing so it is fairly certain (strong) that they will save time (positive/strong) immediately (strong). In other words, instead of teaching motorists to drive more safely, the consequence of ticketing has the effect of teaching them to watch carefully for highway patrol cars.

In Table 2-1, a partial ABC Analysis of these three situations presents an assessment of consequence strength. Smoking, speeding, paying insurance premiums—each is listed with two consequences, one of which is of the

Table 2-1. ABC analysis of consequences for and against

FOR	AGAINST
Paying Insurance Premiums	
Coverage:	Discretionary spending:
later—uncertain—positive	soon—certain—positive
weaker	strongest
Exceeding the Speed Limit	
Saves time:	Traffic citation:
soon—certain—positive	soon—uncertain—negative
strongest	weaker
Cigarette Smoking	
Pleasure:	Lung ailments:
soon—certain—positive	later—certain—negative
strongest	weaker

strongest kind (soon-certain-positive), and the other is weaker, being a mix of strong and weak effects. In each case the stronger consequence favors behavior that common sense may think is either unsafe or short-sighted: continuing to smoke cigarettes, exceeding the speed limit, ignoring the need for insurance coverage. These ABC Analysis sketches from everyday life are similar to safety considerations that arise in industry. Many incentive programs offer rewards that are structured like an insurance policy—the workers are expected to spend their energy or their time on a safety effort (similar to paying premiums) for the sake of an outcome that is positive, but whose power as a consequence is weak because it is deferred and uncertain. One thinks of lottery-like safety competitions where winning is beyond the player's control.

Similarly, many safety programs are oriented toward penalties and punishments, rather like the traffic citation for speeding. Such a program can have some effect in those very rare situations where being caught and cited soon is very certain to follow as a consequence of infringement of the rules. However, even in this case, the management effort would be wasteful because it would be spending its resources on delivering *negative* consequences. Thus even if the negative approach did not have unhelpful side effects, it would be less effective than a positive approach. This is because, dollar for dollar, negative consequences are less powerful in their impact on worker behavior than positive consequences are. However, there are always side effects, and behaviorally trained managers know that the punitive approach presents several serious problems. The usual effect of such a program is to teach people not to get caught. As a practical matter, in most facilities it is as prohibitively expensive to ticket every safety infraction as it is for the police to ticket every driver who exceeds the speed limit.

Finally, a safety effort that relies solely on the threat of some accumulating physical debility to motivate workers is like a stop-smoking campaign that expects smokers to quit smoking just because they have been given the information about associated lung problems. ABC Analysis brings clarity to these matters by insisting on a simple fact: when a facility's safety effort is not working it is because the consequences in favor of safe behavior are weaker than the consequences in favor of unsafe behavior. The function of ABC Analysis therefore is to understand a facility's most stubborn safety problems. (Table 2-2 shows the range of consequences that result from the eight possible combinations of soon-certain-positive.)

One of the most powerful consequences at work in any organization is peer pressure, which offers immediate, certain, rewards to the worker who conforms by accepting him or her as *one of the guys.* A supervisor, too, may inadvertently encourage or maintain unsafe behavior by praising production performance even though the supervisor saw the workers risking accident or

Table 2-2. The range of consequences

STRONGEST	soon	certain	positive
STRONGER	not soon	certain	positive
	soon	not certain	positive
	soon	certain	not positive
WEAKER	not soon	not certain	positive
	soon	not certain	not positive
	not soon	certain	not positive
WEAKEST	not soon	not certain	not positive

injury while they were hustling. The effect of this praise is further compounded when the workers know that the supervisor saw them taking risks. Finally, the most common consequence in favor of unsafe behavior is simply that most unsafe behavior is not even observed, let alone noted and addressed. In this case, not being observed represents a soon-certain-positive consequence in favor of unsafe acts—the worker knows (certain) that unsafe behavior goes unnoticed (non-negative=positive) continually (soon).

ABC Analysis of Failure to wear respiratory protective equipment

Behavioral analysis of antecedents and consequences (ABC Analysis) is a tool that can be used in several areas of behavior-based safety management. First, ABC Analysis illustrates the problem that safety management is up against in most facilities—there are many soon-certain-positive consequences in place that favor *unsafe* behavior. This makes ABC Analysis a good tool for introductory presentations of the behavior-based approach. ABC Analysis is also valuable during accident investigations because it brings a clear understanding of the tangle of consequences that elicited the behavior that precipitated the accident. And third, ABC Analysis serves a similar function during the problem-solving sessions that develop and refine a facility's inventory of critical behaviors.

ABC Analysis has three steps: *Step 1.* Analyze the Unsafe Behavior, *Step 2.* Analyze the Safe Behavior, and *Step 3.* Draft the Action Plan, see Figure 2-4 for an overview. In Step 1. the unsafe behavior under analysis is stated in objective, observable terms, see Figure 2-5. This requirement is very important. From the outset ABC Analysis is focused on behavior, *Failure to wear respiratory protective equipment* not on attitude or blame. This non-judgmental

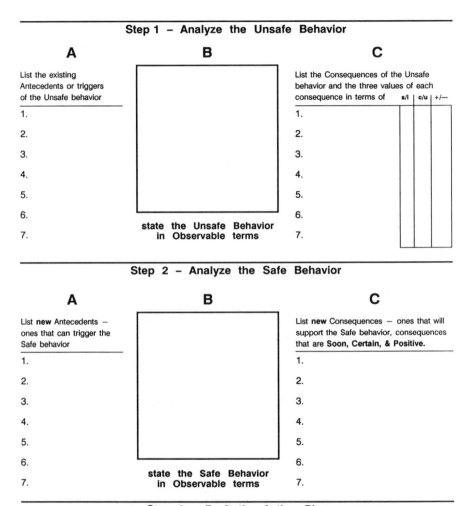

Step 1 – Analyze the Unsafe Behavior

A

List the existing
Antecedents or triggers
of the Unsafe behavior

1.

2.

3.

4.

5.

6.

7.

B

state the Unsafe Behavior
in Observable terms

C

List the Consequences of the Unsafe
behavior and the three values of each
consequence in terms of   s/l | c/u | +/—

1.

2.

3.

4.

5.

6.

7.

Step 2 – Analyze the Safe Behavior

A

List **new** Antecedents –
ones that can trigger the
Safe behavior

1.

2.

3.

4.

5.

6.

7.

B

state the Safe Behavior
in Observable terms

C

List **new** Consequences – ones that will
support the Safe behavior, consequences
that are **Soon, Certain, & Positive.**

1.

2.

3.

4.

5.

6.

7.

Step 3 – Draft the Action Plan

1. Provide new antecedents for the identified behavior/s.

2. Measure the behavior/s by Observation.

3. Apply new consequences systematically (charted & verbal feedback).

Figure 2-4. The 3 steps of ABC Analysis.

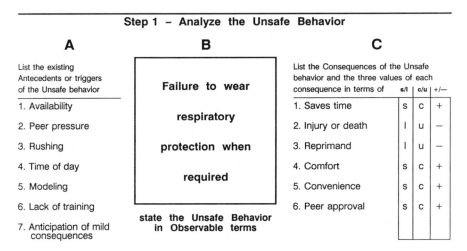

Figure 2-5. Step 1 of ABC Analysis.

approach characterizes the discussions and interviews with workers, and from these a list of the triggers or antecedents of the unsafe behavior is developed. The antecedents represent the things that workers are aware of as triggers of the unsafe behavior. In cases such as this, workers are typically aware that the respiratory protection is unavailable, or perhaps it is available but inconveniently so. Workers are also aware that there is peer pressure not to wear the protection and that when they feel rushed they are much less likely to wear the protective equipment. Workers may say that whether they wear respiratory protective equipment depends on the time of day or shift; and they may note that even supervisors sometimes do not wear protection when it is required, and so are not very good models of safe behavior in this matter. Workers may also say that they have never really been trained to take the protective equipment seriously. And, finally, workers often admit anticipating that no one will reprimand or penalize them when they do not wear respiratory protective equipment. Each of these points is then entered in the following list of antecedents of unsafe behavior:

Antecedents of *Failure to wear respiratory protective equipment:*

- Availability
- Peer pressure
- Rushing
- Time of day
- Modeling
- Lack of training
- Anticipation of mild consequences

The consequences are then listed. These are the consequences that the workers are aware of as following from their behavior when they fail to wear respiratory protective equipment. Under this heading, workers typically note such things as convenience, comfort, or time pressure, saying that it saves time to work without respiratory protection. Workers are aware that they may receive a reprimand and that injury, or even death, may be a consequence of their behavior. However, a worker may also be aware that peer pressure favors a kind of bravado in this matter, rewarding the worker who shows a disregard of respiratory protection. These consequences represent pressures on the worker either for safe behavior or for unsafe behavior:

Consequences of *Failure to wear respiratory protective equipment;*

- Saves time   soon-certain-positive *strongest*
- Injury or death   late-uncertain-negative *weakest*
- Reprimand   late-uncertain-negative *weakest*
- Comfort   soon-certain-positive *strongest*
- Convenience   soon-certain-positive *strongest*
- Peer approval   soon-certain-positive *strongest*

Each consequence that is listed is also analyzed in terms of the soon-certain-positive scale to see which ones are in fact controlling the behavior. In this list, the consequences in favor of the unsafe behavior—saving time, comfort, convenience, and peer approval—are of the strongest kind. Whereas the consequences in favor of safe behavior—possible injury or death, and reprimand—are of the weakest kind. This balance of consequences in favor of unsafe behavior is very representative. It is quite common for the natural consequences in the work place to favor unsafe over safe behavior. Step 2 of ABC Analysis addresses this fact.

Figure 2-6 shows the elements of Step 2. The desired safe behavior is stated in objective, observable terms. New antecedents are listed, ones that can trigger the safe behavior; and new consequences are identified, ones that will strongly support the safe behavior.

Again, it is imperative that the safe behavior be stated in operational terms, terms that avoid reference either to worker attitude or vague injunctions such as *Pay attention, be alert.* There are several reasons for this insistence on objective definition of the safe behavior, not the least of which is that in Step 3 the definition of this safe behavior will form part of the facility's inventory of critical behaviors, to be used by Observers to measure the facility's safety performance baseline and then to track long-term performance. To be useful to the Observers, the statement of the safe behavior has to be in terms of what they can actually see the workers doing.

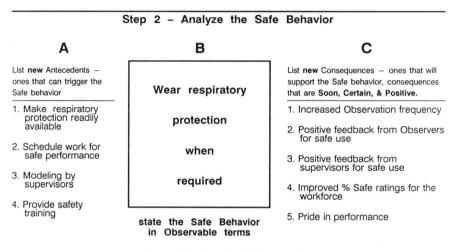

Figure 2-6. Step 2 of ABC Analysis.

The list of new antecedents emerges from a consideration of the old antecedents—availability, peer pressure, rushing, time of day, modeling, lack of training, and anticipation of mild consequences. These are the antecedents which trigger the unsafe behavior, and they need to be addressed. The new antecedents suggest themselves as ways to address the issues raised by the old antecedents. What can be done to make respiratory protection more conveniently available to workers who need it? Can peer pressure be engaged in favor of the safe behavior? Is there a way, through scheduling or otherwise, to manage work that requires respiratory protection so that workers either do not rush while doing it or at least slow down enough to use more readily available protective equipment? A similar question arises about time of day as an antecedent of unsafe behavior. How might supervisors perform as betters models of this behavior? Is safety training indicated for workers? The answers to these questions suggest a list of new antecedents to replace the old ones. However, since antecedents determine behavior to the degree that they predict consequences, the real work of Step 2 is to identify and list new consequences—ones that will support the desired safe behavior. In the search for new consequences, the management challenge is to arrive at consequences that are of the most effective kind—soon, certain, and positive.

Before listing some possible consequences for this particular safe behavior, first a word about the overall suitability of the observation and feedback system to supply the most effective consequences—performance-related feedback. The behavior-based safety process is geared to the challenge, providing systematic feedback on performance and addressing each of the issues of immediacy, certainty, and positiveness. In addition, the observation and feedback process provides measurement through random samples of upstream factors. (For a detailed presentation of measurement considerations see Chapter 3.) In this case the upstream factors are safety-related behaviors and would include such categories as wearing, or failing to wear, respiratory protection when required.

Certain. Using the facility's critical behaviors Data Sheet, trained Observers drawn from the workforce make frequent, sometimes daily, calculations of the percentage of safe and unsafe behavior of the work place. This practice means that workers know for certain that their safety-related behavior is going to be noticed and measured.

Soon. As for how soon the consequences come back, verbal feedback is given immediately—either during or just after observation. The summation and tracking of the data from the Observer Data Sheets gives the workforce daily reports that are focused on its particular safety performance, and the emphasis is on the positive, measuring improvement rates and levels of achievement.

Positive. Many of the other positive aspects of the behavior-based approach have to do with employee involvement—during Data Sheet development, observation, and the problem-solving phases of the process. Where labor unions are present, union representatives take part at all stages of development and Implementation. Hourly employees often make the best Observers—they know both the work and the workforce intimately, and the workers know them. The only name the Observer writes on any Data Sheet is his or her own (and even this identification is optional at some facilities), and for its part management is careful not to use the observation Data Sheets for disciplinary purposes. As a result it becomes clear to the workforce that Observers are not traffic cops intent on citing them for violations.

The outcome of this process is a responsive and statistically valid measurement of the safety performance of a facility, in advance of any injuries or accidents. Since it was developed with input from hourly employees, the behavior-based observation process has credibility with them. Feedback from the measurement process becomes a powerful consequence, therefore. The workforce receives this feedback in the form of posted charts and graphs and in reports at safety meetings. Supervisors also learn that they will be judged rationally and fairly on their safety performance. The measurement process puts in place a mechanism that delivers in a positive way the strongest possible consequences for continuous improvement in safety.

In Step 2 new antecedents are reviewed, and a list of new consequences is composed. These consequences are based on a variety of feedback strategies. In practice, the targeted behavior of this example, *Wearing respi-*

ratory protective equipment when required, would be one of a set of targeted behaviors. Typical new antecedents would be:

1. Make respiratory protection readily available.
2. Schedule work requiring respiratory protection so that compliance is easier.
3. Modeling of safe behavior by supervisors.
4. Provide safety training.

New consequences for the safe behavior would include such items as:

1. Increased observation frequency for this critical behavior.
2. Positive feedback from Observers for safe use.
3. Positive feedback from supervisors for safe use.
4. Improved %Safe ratings for workforce.
5. Pride in performance.
6. Accomplishment.

Step 3 of ABC Analysis consists of drafting an Action Plan. The Action Plan specifies how to provide the new antecedents and the new consequences in terms of observation and feedback. A behavioral Action Plan addressing the analyzed situation would look something like the following example. Note that the Action Plan specifies roles and responsibilities, and deadlines for compliance along with the actions.

Sample Action Plan

1. Assure ready and consistent availability of respiratory protection. Problem solve with Maintenance new locations for receptacles and with Safety personnel for inspection and replenishment schedules. (antecedent intervention)
    — Department Manager, to complete by date: _____.
2. All work groups to view training video on respiratory protective equipment.
                                    (antecedent intervention)
    — Department Manager, training to begin in one week (date: _____)
and to be completed by date: _____.
3. At supervisory and first line safety meeting, management specifies in clear terms its expectation that all supervisory personnel consistently model the safe behavior of wearing respiratory protection where required by facility policy.
                                    (antecedent intervention)
    — Department Manager, meeting to be held date: _____.
4. Observers to begin increased observation frequency for this and related items on the facility's Data Sheet.   (consequent intervention)

—Department Manager, to begin in one week date: _____, and to run for three months, until date: _____.

    5. Positive verbal feedback for satisfactory behavior whenever it is observed.

(consequent intervention)

—Observers and Supervisors, for duration of stepped up observation schedule for this behavior (see dates, Action No. 4).

SUMMARY

Once it has been established, the behavior-based continuous improvement safety process represents a closed loop (see Figure 2-7). During Assessment and Implementation, however, a facility's efforts advance through the stages of the process—identifying critical behaviors (Chapters 6 and 12), problem-solving to develop an Action Plan (Chapters 12 and 13), measuring performance (Chapters 14 and 15), and evaluating for acceptable progress (Chapter 16). After that point, the loop closes with further adjustments of the process, and the facility is on its way to establishing a mechanism for continuous improvement. For instance, if performance does not improve at an acceptable rate, re-analysis of the critical behavior/s is called for. Problem-solving

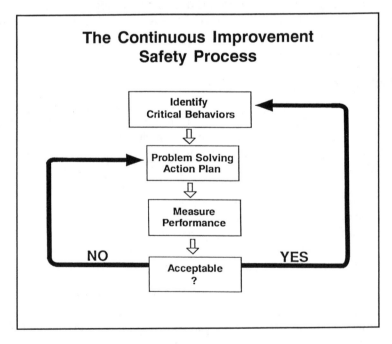

Figure 2-7.

activity then refines the objective definition of the critical behavior in question or reschedules the observation and feedback for the behavior, or both. This process continues until safety performance is acceptable. When safety performance reaches sustained acceptable rates, the process identifies new critical behaviors for problem-solving and measurement.

In addition to this improvement process, various special applications of the behavior-based safety process can be focused on to help employees who have had multiple accidents or to introduce a workforce to the principles of back-injury prevention. In the parts of this book on Assessment, Implementation, and Special Applications the aspects of this process are presented in detail.

Measurement concepts, issues, and techniques form a very important part of behavior-based safety process. The next chapter presents this subject in its relation to other approaches to safety.

Chapter 3

Measuring Safety Performance—
The Process Approach

Good management depends on good measurement. Where human performance is an issue, the difficulty of measurement has been an area of frustration, and therefore innovations of measurement in the behavioral area are welcomed by the concerned manager. The thrust of this book is to recommend a way to measure, and therefore manage, safety performance. This measurement approach deals with safety performance *as a process* and uses a variety of techniques drawn from behavioral and statistical methods to assess the proportion of safe to unsafe behaviors in targeted areas of work place performance. In this Introductory section of the book, the chapter on the *Foundations* of behavior-based safety management provides the larger context for this measurement approach.

Working in the field of behavior-based safety management since 1980, the authors have seen numerous companies in a wide variety of industries take proactive control of their safety performance. In comparison with the predictive power of the behavior-based process approach to safety performance, the reliance on accident frequency as the sole measure of performance shows up for what it is—misleading and reactive. The inaccuracy of accident frequency numbers as the sole measure of performance is that an accident is an event, a discrete thing; whereas safety performance is an ongoing process consisting of thousands, or even hundreds of thousands, of safety-related behaviors. In the absence of process measurement addressed to behavior, it was understandable that people would focus on accidents. If almost nothing else is simple about accidents, at least they are easy to count, and clearly accident frequency is an important thing to know.

It is not nearly so clear, however, just what is indicated by the accident frequency number for a given facility. By the definition of an accident, when it occurs one knows that something adverse and unfortunate has happened. But does one know how likely it is to happen again? Something should be done but what is to be done to prevent the kind of accidents that account for the frequency numbers? Accident frequency is not a measure that addresses these and similar questions. Furthermore, when accident frequency is used as though it were the sole measure of safety performance it is a source of confusion and misguided effort. One example is the common practice of "managing the numbers." Injuries are reclassified into categories that make the safety effort look good—at the substantial loss of management credibility with the workforce.

This chapter addresses the measurement issues involved in managing real safety performance, not just the numbers. The most valid method of achieving sustainable, long-term results is to steer a facility's safety efforts by a variety of behavior-based indicators, in judicious combination with accident frequency. The presentation of this method which follows draws the analogy between it and the quality improvement process. The discussion of these matters touches on the following subjects:

- Objectives of safety performance measurement
- Upstream safety factors
- Common management errors
- The proper use of accident frequency rates
- Managing the continuous improvement safety process
- Five indicators of safety performance

OBJECTIVES OF SAFETY PERFORMANCE MEASUREMENT

The primary objective of measuring safety performance is to provide a feedback mechanism that will foster improvement—and continuous improvement, at that. The crucial thing about a feedback mechanism is that its effectiveness is directly dependent upon tapping the right sources of information in the first place. Although this principle sounds self-evident, it is easily lost in practice. The most common management mistake about information sources is the reliance on accident frequency rates as sole indicators of performance. This central problem is approached in various ways in this chapter. Not only is the corrective feedback safety mechanism compromised by tapping the wrong data; the secondary objectives of measurement go astray also. Taken as a cluster of overlapping provisions for safety, these secondary functions are also very important, having to do with problem identification, preventive action, and documentation and reinforcement of performance.

Identifying problem areas. Since the reporting forms and procedures in standard use have not been drafted for behavior-based accident prevention, without some training in the behavioral components of accidents, it is hard for safety personnel to identify problem areas even from otherwise conscientious reports. The safety mechanism that is geared for performance measurement and feedback therefore takes its cue from the facility inventory of critical behaviors as developed and focused during Implementation. The same holds true for accident investigation forms and procedures. In this way the crucial phases of information gathering all serve to enrich the facility's inventory of critical safety-related behaviors. In the absence of this integrated approach to identifying facility problems, however, the gathered information contains much that is not useful.

Stimulating preventive action. Without clear information on the relevant problem areas and their contributive upstream factors, accident prevention itself happens by accident. A general area is saturated with management attention and concern, and that effort pays off. This shotgun approach wears people down, and the result is that, whatever they may say, in their own minds managers give up on accident prevention as something that is too expensive for them to afford.

Documenting safety efforts. Added to the preceding situation is the difficulty many facilities have in documenting the effectiveness of their safety efforts. Without accurate documentation there is no true safety accountability rating for managers and supervisors. Nor is there a developing record of facility safety performance trends and profiles—the kind of record that is taken for granted in every other facet of management practice.

Reinforcing improvements in performance. Finally, the strongest reinforcement for behavior is feedback that is soon, certain, and positive (see Chapter 2). It is not enough to know this, however. Managers who want to reinforce improvement must also know precisely which behaviors represent improvements in the safety performance at their facilities. Left to good intentions and hunches on this subject, management continues to reinforce numerous unsafe behaviors.

Problem identification, preventive action, and documentation and reinforcement of performance—these four secondary objectives of safety performance measurement are closely related to each other and they point to the first principle of continuous safety improvement, the principle of process measurement. This type of measurement is already familiar to many managers from its application in the field of quality improvement.

QUALITY AND SAFETY—TWO SIDES OF THE SAME COIN

Behavior-based safety management and quality improvement are closely related, perhaps so closely that they are essentially the same thing. Quality efforts aim to minimize the variability of product qualities. Safety management minimizes the frequency and severity of unplanned and untoward events that harm people. The basic principles at work in the quality improvement process are also at work in the behavior-based safety management process. This is because both of these management processes are based on scientific methodology. Eight of the quality improvement concepts that apply directly to behavior-based accident prevention are:

1. Constancy of purpose—develop long-term strategies and stick to them.
2. Process, not program.
3. Do it right the first time.
4. Do not blame the employees.
5. Specify standards in operational terms.

6. Use measurement of upstream factors to assess performance.
7. Improve the process, not the downstream results.
8. Use statistical techniques to distinguish variation due to common cause from variation due to special cause.

Concepts 1 and 2 refer to the same thing from different sides. Over the years managers have approached safety as a series of short-term programs. The effect of this practice has been a loss of credibility with the workforce concerning constancy of purpose. What is needed in safety is a process of continuous improvement that does not start and stop. Taking the necessary steps to make sure that things are done correctly the first time, *Concept 3,* applies clearly in both quality and safety. The opposite of *Concept 4,* blaming employees for their injuries, is a very common management error in safety, discussed at greater length below. *Concepts 5 through 8* represent the application of statistical methodology to these two management areas of quality and safety. Although Concept 5 may not seem to be statistical at first glance, it is essential in the development of an inventory of recordable behaviors, the first step toward making a process measurable.

Process approaches to quality and safety are so complementary that when they are both at work in a facility, they tend to reinforce each other, the safety effort gaining strength from the quality effort, and vice versa. On the other hand, the close match between these two methods also means that a facility using the process approach in either safety or quality, but not both, runs the risk of sending an inconsistent message to the workforce, thereby losing effectiveness from the process approach where it is working.

Upstream Safety Factors

In most areas of performance, activities both precede and produce outcomes. Common instances of this relation are found wherever the practice is encouraged as the prerequisite of skill (practice makes perfect), or study as the guarantor of good school grades, or research and development as the source of new products, or hard work as the route to better pay. In the terms of statistical process control (SPC), the processes (activities) are *upstream* (practice, study, research, hard work), and their results are *downstream* (skill, good grades, new product, better pay).

In the area of diet, metabolism and weight management, for instance, it is known that weight gain and weight loss are downstream factors. An athlete and his or her physician or coach will have some optimal weight in mind, but they now realize that the factors to manage are the upstream factors of caloric intake, fat consumption, and energy output. Consequently, although they have targeted some specific weight for the competitor—either as weight gain or loss or maintenance, they are not distracted by short-term fluctuations in the athlete's weight. Similarly in the case of statistically based quality improvement, management looks not at product defects (downstream factors) but at upstream factors of production which are predictive of defects. In the terms of the behavior-based accident prevention process, accident frequency rates represent downstream indicators. Accident prevention relies instead on sampling the mass of safety-related behaviors which lie upstream and which precede any particular incident.

This basic fact about safety bears repeating. A great deal of exposure to risk and hazard has already occurred before any given injury or accident occurs. Each particular incident is precipitated out of a mass of preceding unsafe behaviors (Chapter 2). The purpose here is to highlight this peculiar relationship in safety performance between the mass of upstream factors and the individual accident downstream. Furthermore, the vast majority of exposures occur with no noticeable downstream results. In other words, there is a *demonstrable but indirect* relation between the frequency of unsafe behavior upstream and the frequency of accidents downstream. The indirectness of the relation makes it confusing and hard to grasp, let alone to pursue long-term in an undistracted way. It is this basic fact which causes most of the problems in poorly focused and ineffective safety efforts. Because the linkage is so indirect between the upstream and downstream factors of safety performance, management does not know what to pay attention to and, therefore, tends to overreact to random variability in accident rates.

Common Management Errors

Another way of expressing the difficulties which safety presents to downstream management efforts is shown in Figure 3-1. Cultural factors give rise to various aspects of management systems. Management systems in turn either create or eliminate exposure to risk and hazards. Finally, emerging from the patterns of exposure there are incident rates. Where in this overall system should measurement occur? The old approach focuses measurement exclusively on the end point of the process, and the resultant measure is the accident frequency rate. However, this is at best a limited indicator of real performance, and it provides no information at all about upstream factors such as exposure, management systems, or culture. This is one of the central points of this book.

Suppose that accident frequency is down for one quarter. Does this mean that the whole system should be given a clean bill of health? Practices and procedures that are prejudicial to sustained high safety performance may still be present in the system, and these must be considered along with incident rates. To ignore this fact and proceed on the basis of incident rates alone is analogous to defining the physical health of an individual as the absence of disease. This is not good diagnostic practice even when the

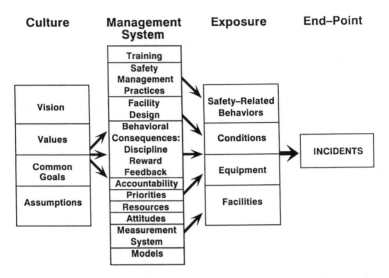

Figure 3-1. Safety incidents are downstream—not upstream.

doctor as health manager does not yet know much about the patient. It is especially poor practice when the physician does know that the individual smokes cigarettes, is overweight, does not exercise, has elevated blood pressure, eats poorly, and suffers from high stress. Managers, supervisors, and hourly employees almost always know it when their facility has bad safety habits. Though they often disagree about the precise nature of the safety problem, they know when they are receiving undeservedly high grades for low accident rates that they have not earned. The limitations of accident frequency as a measurement tool are discussed later.

Along with confusion about accident frequency, the other common error of management is to blame the employee for accidents which are perceived as arising from "stupid" actions of employees. This reaction, as understandable as it may be, is a mistake on two counts. In the most basic human terms it is counter productive to blame employees as though they got hurt intentionally. This approach puts the employee in the position of resentment and resistance. As important to management effectiveness, blame is also a mistake because it fails to take into account how the overall management system influences employee behavior. The manager versed in behavior-based techniques knows for certain that when an injury results from an unsafe behavior, the implicated unsafe behavior has been performed on numerous previous occasions. This is another aspect of the linkage described above—many exposures occur prior to any injury. Therefore the responsibility for this behavior and its frequency rests with the management system as well as with the employee.

For instance, when an employee is injured while reaching into a moving line to clear a jam, it is tempting for the frustrated supervisor to react in terms such as, *What's wrong with my people? These guys are stupid!* However, the underlying and very pertinent question is: what consequences are in place that *favor* the action of reaching into moving machinery? Knowing that supervision has either implicitly or explicitly condoned this behavior in the past, the investigative question becomes, *why is the behavior allowed to continue?* Employees are justifiably resentful in cases where their unsafe behavior has been either encouraged or allowed until an injury occurs, and then they are not only injured but subjected to disciplinary action. It is the responsibility of management to establish and maintain systems that produce safe behavior and discourage unsafe behavior.

THE PROPER USE OF ACCIDENT FREQUENCY RATES

The standard computation for accident frequency rates is

$$\frac{(\# \text{ of accidents}) \times (200{,}000)}{\text{total hours worked}}$$

This calculation gives a figure for the number of accidents per 100 employees per 12-month work period. Although the calculation is straightforward and in wide use throughout industry, it is nonetheless easy for managers to lose sight of what these frequency rates actually mean for their safety efforts. When a period of time passes with no injuries, it is tempting to think that safety performance is good or even improving. Conversely, when several accidents occur within a short period, it is easy to conclude that performance must be down or declining. Neither of these need be true. The true significance of either of these results can only be determined statistically, by comparing these outcomes to the predicted random outcome for the facility in question. A stable system will produce a variable number of injury events.

Take the example of a work group of 100 employees that has just completed the first quarter of their current safety measurement year. For the preceding 12-month period, the work group has a recordable rate of 2.0. In the 3-month period just completed, however, the recordable rate rose to 4.0. At first glance it appears that safety performance has declined significantly, but what the quarterly number actually indicates is that one injury has occurred within the past three-month period. Given the work group size, the apparent quarterly increase to 4.0 does not represent significant variability (special cause). Although taking account of random variability is standard procedure using statistical process control in the quality improvement process, an awareness of random variability remains rare in safety management. In the field of safety many reward systems and performance appraisals are based on numerical goals and measures that are untested for random variability. For the supervisors of this hypothetical work group, this could well mean receiving a bad performance rating that is undeserved.

On the other hand, the unclearness of this concept in safety also means that work groups often get a good safety rating when they do not deserve it either. For instance, suppose a work group with the same past-year frequency rate of 2.0, but a first quarter in which no recordable injuries have occurred for a 3-month frequency of zero. This apparent improvement in their safety performance is also just that, merely apparent. For a work group of 100 with an annual rate of 2.0, a quarterly range of frequencies from 0.0 to 4.0 has *no statistical significance.* This range of random variability of outcome is typical of a stable system in which there is *no change* in the ratio of safe to unsafe behaviors.

As another example, Figure 3-2, take an organization with 1000 employees and a recordable incident rate of 2.0. The numbers for this hypothetical case describe a facility in which 20 injuries are expected in a 12-month period or fewer than 2 per month. This is an extremely low number of events for a population of one thousand people exposed daily for a 1-year period. On

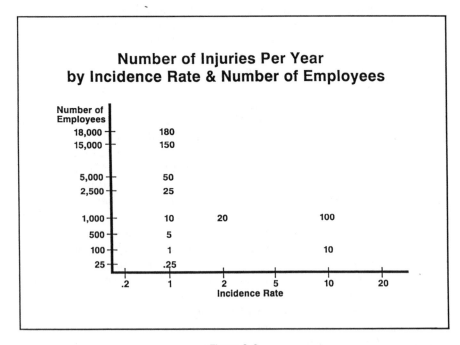

Figure 3-2.

average at this frequency rate in this population, an individual would experience 1 injury for every 50 years that he worked. This is another way of saying that the rate is 1 injury per 50 man-years of work. Because accidents at this frequency level are such rare events, a very large number of hours worked must be logged before this method of measurement achieves statistical validity. How many man-hours are needed for statistical validity? Approximately one million man-hours are needed in order to establish a lower control limit for a frequency rate of 2.0. In other words, a facility whose injury frequency rate is 2.0 needs to log approximately one million injury-free man-hours before it can be certain that they have achieved improvement and not just randomly low numbers of incidents. Figure 3-3 graphs the minimum time required for work groups of varying sizes but all with a frequency rate of 2.0.

For the typical facility these laws of statistics mean that the injury frequency rate is a measure that is accumulating validity as time goes by but that these frequencies are of no predictive value to safety management on a monthly or even quarterly basis, let alone a weekly or day-to-day basis. On the contrary, given the misplaced trust that people accord to frequency rates, these numbers are an outright hindrance to proactive safety management.

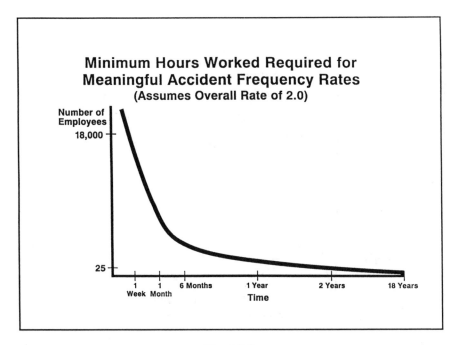

Figure 3-3.

Reliance on such incomplete statistics as the primary indicator of safety performance gives rise to the accident cycle (see Figure 3-4).

This graph of the accident cycle shows the relationship between a facility's thresholds of acceptable injury rates and management's changing response to the rates. When the recordable rate goes above a facility's upper limit, management acts to drive the rate down. When the rate falls below the acceptability limit, management loses focus on safety, and the recordable rate rises again. In this cycle the management action for improvement follows fluctuations in the accident frequency.

This means that companies who respond to the safety cycle and work very diligently to push down their accident rates, make a new problem for themselves. Since they are triggering their intervention off of a downstream measure, they lose their focus when rates are low. They may want to work on safety because they know that when they do not, their accident rates will rise again. Some high-performing companies beat the safety cycle by working on safety no matter how low their accident rates go. However, although their rates no longer cycle, they usually reach a point beyond which they do not improve, either. By achieving a low but unfortunately stable rate, the com-

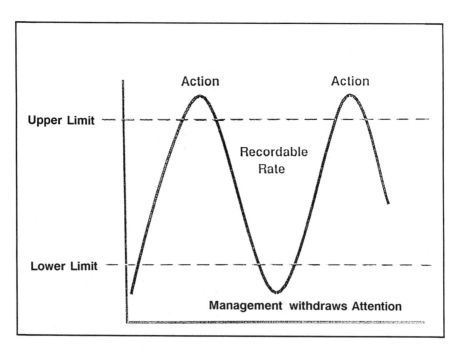

Figure 3-4. The Accident/Safety Cycle.

pany beats the safety cycle only to stall on a safety performance plateau. Additional effort achieves very little added benefit.

In either case then—either safety cycle or safety plateau—the strategy of sole reliance on accident frequency means that continuous improvement is not possible, because for companies unaware of the safety cycle, improvement efforts are contingent upon cycles of poor performance. Whereas companies that do know better than to let their safety efforts be driven by the safety cycle, nonetheless often do not know what else to steer by and so their improvement levels off and stalls.

In addition to these drawbacks, when incentives such as prizes, awards, and privileges are based on accident frequency, false feedback is the outcome. Suppose that work group *A* has done a poor job on real safety performance, but since they were "lucky" in the random variability of their rate and had no injuries for the period in question, they receive the incentive. In effect they are rewarded for their poor performance. Suppose that work group *B,* on the other hand, did excellent work on real safety issues, but since they have had an injury their performance goes unrewarded. In effect they are punished for their good performance. This system of reward is a lottery of sorts, one which strongly encourages managers to make the numbers look good. Managing the numbers can be done by reclassifying injuries, creating special categories of rehabilitation, influencing medical treatment, and so on. The net effect is a waste of time and resources, one which furthermore erodes the credibility of the organization's commitment to safety.

It is very important to be clear that real safety performance is never a matter of luck. Of course random variability (chance, luck) is present in workforce activity, just as it is present in any statistically complex chain of events. However, *managing* safety performance has nothing to do with chance in the sense of "getting lucky." No, managing safety performance has to do with *chance* in the technical, statistical meaning of the word—genuine safety management acts to limit the range and seriousness of the effects of chance or random variations in workforce behavior and work place conditions. Limiting the effects of chance is a key management function. To say of good managers, they leave nothing to chance, means that whatever their field of responsibility, good managers limit random variation to those areas and thresholds that do not threaten the enterprise they manage, they do not leave the incidence of safety-related behaviors to chance.

Summarizing, there are six important negative effects of using accident data as a primary indicator of safety performance:

1. This approach is reactive rather than proactive.
2. Random variability is misread.
3. As a consequence of negative effect #2, management overreacts.

4. Safety incentives based on frequency rates amount to false feedback.
5. The emphasis on frequency rates encourages mere numbers management.
6. The net result of 1-5 is an erosion of the credibility of the safety effort.

In contrast to this flawed and frustrating approach to safety, the continuous improvement safety process is based on the measurement of safety-related behaviors rather than of incidents.

THE CONTINUOUS IMPROVEMENT SAFETY PROCESS

The steps of the continuous improvement safety process are the same as in the continuous quality improvement process: specify standards, measure compliance, and provide feedback on improvement. While these points seem straightforward, the difficulty comes in the application of them to safety. What is the most effective way to specify safety standards? Just what is measured when compliance with the safety standards is measured? On what basis is feedback provided on performance, and to whom and how often?

For safety, the first step of specifying standards is accomplished by developing an inventory of critical safety-related behaviors and defining them in operational terms. To measure compliance, trained Observers use the inventory of behaviors to rate the facility, producing a read-out in terms of the ratio of the safe to unsafe critical behaviors. Providing feedback on improvement then amounts to giving the workforce charts and reports of their progress on the inventory of behaviors. See Figure 2-7 in the preceding chapter for an overview of the data flow in the continuous improvement safety process. Looking again at the overall process that produces injuries (review Figure 3-1), it becomes clear that the proper place to measure safety performance is at the level of exposure. In the data flow figure (see Fig. 16-1) this level corresponds to the Observer making his or her notes on workers doing a job. Statistically valid measurement of exposures is more suited to the typical facility because there are so many more safety-related behaviors to sample than there are injury incidents. In effect, once these safety-related behaviors are identified and defined operationally, they become critical behaviors. In this way the safety effort becomes a mechanism for continuous improvement, guiding itself by the incidence of unsafe behaviors rather than by injuries and regularly updating and focusing its inventory of critical behaviors to match developing safety issues. Also, the analysis of persistent unsafe behaviors highlights management system issues, and problem-solving addressed to such issues drives the company culture toward stronger safety performance.

Five Indicators

The following is a summary of five indicators: accident frequency, frequency of observation, exposure levels or %Safe readings, safety-related maintenance information, and involvement indicators and surveys.

In a number of informal studies, frequency of observations per 100 employees has been shown to be a consistent predictor of accident frequency rates. An inverse relation holds here—frequency of observations up, accidents down, and vice versa. The advantage of rate of observations as a measure is that feedback about it is available on a much more frequent basis than information about accident rates, and this allows management to employ preventive steps.

The %Safe figures produced by observation are also good indicators; however, management must be very careful not to put pressure on this indicator. This indicator is soft by nature, and employees must not be overly motivated to look good on this measure. Various involvement indicators can be quantified —such measures as frequency of attendance at safety meetings, number of employes trained to be Observers, and activity levels of safety committee meetings on specific projects. Response to safety-related maintenance is also an easy area to quantify and track as an overall indicator of performance. Even though safety-related maintenance may not be the cause of a high percentage of injuries, such information displayed publicly increases the credibility of management's overall commitment to safety.

Finally, although more research is needed to define their relationship with accident frequency, informal studies suggest that standardized safety climate surveys are also predictive. They could be given semi-annually.

Management decides how often the different levels of the organization receive these five indicators (accident frequency, frequency of observation, exposure levels or %Safe behavior, safety-related maintenance information, and involvement indicators and surveys). Table 3-1 shows a possible preliminary schedule for these indicators for various typical employee levels. In a very short time, the work group and first-line supervisors learn to make good use of exposure indicators or %Safe data on a daily basis. This indicator is itself daily and it gives proactive, immediate, positive and corrective feedback for identifiable areas of safety exposure. These exposure indicators make immediate sense to both the worker and the supervisors. This same level may want to see its observation frequency data on a weekly basis. The same weekly information is often helpful to site managers and senior managers, along with a weekly summary of exposure data.

On a monthly basis, safety-related maintenance indicators are helpful to all levels of the organization. On a quarterly basis it can be helpful for site managers to review accident frequencies, adding involvement indicators and

Table 3-1. Target feedback by frequency and employee level

	DAILY	WEEKLY	MONTHLY	QUARTERLY	SEMI-ANNUALLY	ANNUALLY
Workgroup and First-Line Supervisors	Exposure	Observation Frequency	Safety-Related Maintenance		Involvement Indicators and Surveys	Accident Frequency
Site Managers		Observation Frequency, Exposure	Safety-Related Maintenance	Accident Frequency	Accident Frequency Involvement Indicators and Surveys	
Senior Managers		Observation Frequency, Exposure	Safety-Related Maintenance	Accident Frequency		Involvement Indicators and Surveys

surveys semi-annually. Work groups and supervisors can also profit from a semi-annual review of involvement indicators and surveys, and from an annual review of accident frequency.

SUMMARY

Management and measurement go together. The adaptation of this measurement-oriented safety process begins with Assessment at a facility. From a measurement perspective, the challenge during Assessment is to employ behavioral analysis of accident reports and statistical analysis of injury data to extract categories and frequency rates that the behavior-based approach can work with. This task can be difficult at first, because most accident report forms and procedures are not concerned to document the behavioral components of an injury. Similarly, it is rare to find injury data that has been formatted with subsequent observation and measurement in mind. Nonetheless, in fairly short order, a small committee of knowledgeable people can achieve genuine gains in clarity and understanding of these matters. Their findings, communicated in the Assessment Report (Chapter 8), identify the cluster of behavioral categories associated with the majority of the facility's injuries. They may also indicate in which departments/shifts it would be most fruitful to concentrate the behavioral measurement process.

Assessment also accomplishes two other important objectives. Because it gets input from a broad cross section of a facility's personnel, the Assessment effort also serves as a training opportunity, introducing a significant proportion of the workforce to the behavior-based approach to continuous improvement in safety performance. At the same time, through interviews and surveys, the Assessment identifies those aspects of the safety culture which favor Implementation of the behavior-based approach, and those aspects where resistance is to be expected. A clear assignment of roles and responsibilities is outlined (Chapter 9), and key members of the Implementation receive training in how to manage resistance to change (Chapter 11).

PART 2. ASSESSMENT

Chapter 4

Introducing the Behavior-based Accident Prevention Process to a Facility

One approach to a safety initiative is to *bring in the new and throw out the old*. This approach is ill suited to Implementing the behavior-based accident prevention process, however. The abrupt attitude of *throw out the old* is one that loses credibility with hourly employees, whereas Implementing the accident prevention process builds employee trust in the mechanism, enhances their involvement, and creates in them a sense of ownership of the process. This does not mean, *Turn safety over to the workers.* It means a balance of management leadership and employee involvement.

The roles and responsibilities of leadership consist in showing where opportunities and resources exist, in setting the direction of the process, in defining the long-range objectives of the safety effort, and in establishing a supportive environment—one that drives out fear, encourages participation, and rewards good performance. Employee involvement, however, is the very mechanism of *continuous* improvement. Hourly employees come to understand the objective. They know why it is essential, what it means in daily activities, and they have a sense of ownership. In short, they become stakeholders. Safety is an ideal place to start developing these valuable qualities in a workforce. Even in facilities with poor employee relations, safety is an area that represents common ground—everyone benefits directly and visibly from improved safety performance.

The patterns of success in these matters are clear. Facilities that achieve continuous improvement as a safety goal do so by following certain steps and by treating each step as an opportunity to involve hourly employees; for an overview, see Figure 4-1. During a facility's Assessment phase of the safety process, while the Assessment team is learning about the safety culture there and planning a successful Implementation they are also fostering the employee relations just described. The Assessment interviews and surveys are used as opportunities to inform employees about the continuous improvement safety process. Two other important resources for employee involvement during this phase are training seminars and on-site introductory presentations.

Initial Sponsorship

The sponsorship of the Assessment phase varies from facility to facility. Initiative may come from any number of key people. Sometimes it comes

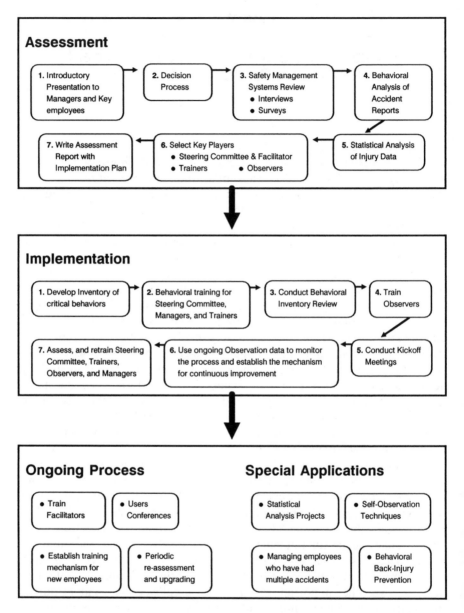

Figure 4-1. Establishing the safety mechanism—an overview.

from the site manager with one or two interested department heads, or it may come from the safety and/or quality department heads with the input of key supervisory and hourly employees. A corporate officer may start the ball rolling at a particular site. Union leadership sometimes takes a sponsoring interest in the behavior-based approach, seeing in it a way to address work place safety that steers clear of personality clashes and the like. This informal group usually includes three to five key people, and however this group is composed, its formation is a good sign. In most cases the appearance of this group is predictive of a successful Implementation effort. Typically, the informal group members discover their mutual interest in the behavioral approach after an introductory presentation of it and its foundation concepts and before Assessment is underway. They may have attended a training seminar together, or an on-site presentation. What counts is that the members of the informal group are enthusiastic about the behavior-based safety process, and they convey their enthusiasm to other key people, meeting with them, taking them to dinner, persuading them to at least undertake a thoroughgoing Assessment. At the same time that this group is lobbying for the behavior-based approach, they oftentimes also set up the facility's Steering Committee, discussing candidates for Committee membership, approaching them to sound them out, and even arranging for training for them. (See Chapter 9 for a detailed presentation of Steering Committee issues.) Training is important throughout the behavior-based approach to safety improvement, and the most effective Assessment and Implementation are the ones that capitalize on the training opportunities present at each step of the way.

Using Available Training Resources

The best way to approach the safety process is to involve employees in the very first investigative steps of the process. Forward looking management includes hourly employees on the facility evaluation team and sends the team to visit plants using the behavioral process, and to training seminars in behavior-based methods. The hourly members of the team are asked to provide their input to such vital questions as:

Will this approach work at the facility?
What obstacles to this approach might exist at the facility?
What would the facility need to do to assure long-term success of this approach and enlist widespread support for it?

The Introductory chapters of this book on *Measurement* and on *Foundations* are training resources also.

On-Site Introductory Presentations

On-site presentations of the accident prevention process are another important early step. These presentations are often conducted by an outside consultant. However, effective introductory presentations may also be made by the facility's own representatives, or by a knowledgeable person from another facility or from corporate headquarters. In any case hourly employees attend these meetings. Inclusion of hourly employees at the introductory meetings accomplishes several important things.

- A valuable source of facility input is assured from the outset.
- Employees realize that they are part of the process *from the beginning.*
- Trust is formed.

Then, in the unusual case where hourly employees are negative about the process, the important thing is to listen to the issues that they raise and to address them. This strategy, pursued from the outset, is the best assurance of success.

After the accident prevention process has been evaluated by a representative cross section of a facility's employees, the ground has been properly prepared for the Assessment phases discussed in the remaining chapters of this part of the book.

ASSESSMENT—AN OVERVIEW

The factors that distinguish a facility with excellent performance in safety from others are work practices, safety culture, and attitude. Work practices are the final common pathway to most injuries. Work practices are behaviors or Observable acts. As behaviors, work practices are triggered by antecedents and reinforced by consequences. Whether workers act safely or unsafely at a given moment is determined by the range of the consequences impinging on them at that time. Attitude and culture are the sources of both antecedents and consequences. An individual's attitude is his or her internal guide to what the culture rewards and punishes. Culture is the set of values and practices that are held in common by a group of people. Culture is *"how we do things around here."* It is *"what we think is important here."* Culture is the external source of the antecedents and consequences that impinge on the worker.

Safety's basic building blocks. A facility's safety building blocks are the antecedents and consequences which have significant impact on safety-related behavior at any given moment. Thus the characteristics which distinguish safer from less safe facilities are present in the culture of the facility and in the attitudes and behaviors of employees at all levels. To grasp the facility's basic safety building blocks, it is necessary to grasp the things in a plant's culture that have an impact on the safety-related behavior of the workers.

Therefore the focus of behavior-based Assessment is to study the safety culture at a facility, engage employees in the improvement process, and on the basis of findings and input from employees to develop an Implementation Plan. Since culture is a dynamic thing that is always changing or moving in some direction, an important question for Assessment has to do with the direction of movement. Is the culture of a given facility moving toward a stronger safety infrastructure or a weaker one? Is the facility consuming its safety capital or adding to it?

As was mentioned in preceding chapters, although the principles of the behavior-based approach to accident prevention are straightforward, their successful Implementation is not. Successful Implementation requires some organizational development, and therefore the purpose of Assessment is three-fold:

- Discovering cultural forces
- Fostering employee involvement
- Addressing planning issues

Discovering cultural forces.   Before Implementing the behavioral process at a given facility it is necessary to have a clear understanding of the cultural forces in place that will assist the Implementation effort and the forces that will work against it. All safety cultures contain elements or factors that will encourage Implementation as well as factors that will discourage it. Successful Implementation of the behavioral process requires a thorough understanding of these competing forces. Assessment is designed to locate these forces in the culture.

Fostering employee involvement.   The second purpose of Assessment is to foster employee involvement in the Implementation effort itself. During the Assessment there are numerous opportunities for providing information as well as collecting it, giving the behavior-based safety process wide exposure and gathering input that is timely and positive.

Addressing planning issues.   Assessment develops answers to planning issues such as:

- How will the cultural issues identified in the Assessment be addressed?
- How will employee involvement be assured?
- What is the Implementation Plan?

- Should the facility start off with a long and comprehensive inventory of critical behaviors or a more focused inventory? Should the facility inventory emphasize generic behaviors, job-specific items, or both?
- Within the organization, who should develop the inventory of critical behaviors?
- Should each department have its own inventory, or should there be one inventory for the entire facility?
- Should Implementation be plant wide or in a limited area? If Implementation is plant wide, should it start with the entire plant or proceed in stages?
- Within the ongoing behavioral process, what are the roles and responsibilities of the first-line supervisors, the work groups, the middle managers, and senior managers?
- Should Implementation be administered through a Steering Committee? If so:
  What is the Committee's charter?
  Who should serve on it?
  Who should be its Facilitator?
  To whom does the Committee report?
  What is the best way to assure the follow-up of Committee recommendations?
- Will members of the Steering Committee be trainers, or will someone else do the training, or both?
- How often do behavioral observations/samples need to be done? Who will do them? Are there hot spots in the organization where observations should be done frequently?
- What is the data flow for the observation Data Sheets? And if computer software is used to process the observation Data Sheets, who will enter the data and produce the computer generated reports, graphs and charts?
- How and when will success of the process be measured?

Assessment Methods

The Assessment effort proceeds along four avenues, each of which plays a crucial role in achieving an accurate picture of a facility:

- Interviews
- Safety Survey
- Behavioral analysis of accident reports
- Statistical analysis of injury data

Interviews. Group and individual interviews are conducted with a number of employees across levels and locations. The purpose of the interviews is to elicit perceptions of safety issues, and descriptions of current safety procedures and management systems to assess the safety culture. In addition, the interviews give employees a voice in how to implement the process most successfully.

Safety Survey. Since surveys are completely anonymous, they provide a more standardized measure, without the effect of peer pressure, to use as a cross-check for the findings of the interviews. In order to provide this important cross-check, the safety survey must be structured to provide information about the key variables of the safety culture of the facility. See Chapter 5 for a presentation of the Assessment Interviews and the Safety Survey.

Behavioral analysis of accident reports. A sample of the facility's injury reports is analyzed to determine behavioral factors that contribute to injuries. This analysis lays the foundation of the facility's inventory of critical safety-related behaviors. This preliminary analysis of accident reports during the Assessment effort identifies the behavioral categories that are most often associated with the facility's accidents and injuries. These same findings guide the Steering Committee during the Implementation phase while they develop the facility's inventory of critical behaviors and its observation Data Sheet. Chapter 6 presents this material in greater detail.

Statistical analysis of injury data. Complementing the behavioral analysis of accident reports, available injury data is reviewed using statistical methods to determine where safety efforts may be most effectively focused. This analysis identifies important variables such as Shift, Department, etc. See Chapter 7 for a sketch of this type of statistical analysis.

The Effective Assessment Report

The effective Assessment Report achieves two important things. First, it sums up, interprets, and presents the results of the four Assessment methods previously listed. Second, it proposes a plan for the Implementation effort to follow. Since Implementation requires that the facility adapt the principles of the behavioral safety process to its own organizational structure, the Assessment Report must spell out the Roles and Responsibilities of the facility's personnel to the steps of the Implementation effort. Chapter 8 presents the Assessment Report, and Chapter 9 addresses some of the important issues that arise concerning Roles and Responsibilities.

Chapter 5

Interviews and Surveys—
Assessing the Safety Culture

INTERVIEWING

Interviews are conducted with all managers and supervisors and with at least 10% of the hourly workforce. These interviews are conducted anonymously, but they are not confidential; material is written down and included in the Assessment Report. The interviewees are not named, and this makes it easier for them to give accurate expression to their perceptions. Where the workforce is unionized, union leadership is included. In facilities with fewer than 500 employees, the proportion of hourly workers interviewed must be higher than 10%, see Figure 5-1. In any case, the interviewing must be sufficiently cross sectional to provide a representative sample of hourly workers from all areas and shifts. (*Note:* when safety surveys are used to supplement interview results, the surveys are administered at the beginning of each interview session—see *The Written Safety Survey.*)

The interviews are leader-directed, and they are held to a fixed time limit. The interview format with senior managers is one-to-one. First-line supervisors and hourly workers are interviewed separately from each other in groups of 8 to 12 people at a time. The group interviews require 1½ to 2 hours to complete. Interviews with senior managers require 1 hour each. These estimates include time for the participants to fill out a written safety survey.

The perceptions that people have about things can be more important determinants of behavior than the realities. For instance, management might be committed to improving safety and may be doing a number of things that are effective from an objective point of view, but if workers believe that management does not really care, the workers are less likely to behave safely themselves. To find out what the issues really are, people from all levels in a facility need to be interviewed—facility manager, his direct reports, middle managers, first-line supervisors, and hourly workers. The people interviewed must be as representative of the general workforce as possible. This means that they should come from all departments, shifts, job categories, and length of service. The interviewees should include both men and women and the major racial or ethnic groups at the facility.

A high proportion of first-line supervisors must be interviewed; in smaller facilities this may mean most of them. A minimum of 20 people is desirable,

Interview Sample Size	
# of employees	# of interviewees
fewer than 50	50 %
50	25 employees
100	30 "
200	35 "
300	40 "
400	45 "
500	50 "
more than 500	10 %

Figure 5-1.

selected to be representative of the supervisory population. The facility manager should be interviewed, as well as all of the people who report directly to the facility manager. The number of middle managers who should be interviewed varies with the size and organizational structure of the facility. If there are relatively few middle managers, they should all be interviewed, and generally speaking it is desirable to interview a high percentage of them.

In union-represented facilities, consideration is given to how the union will be involved, if at all, in the Assessment effort. Site managers may choose to have union representatives from the facility interviewed as a separate group, or the union may select active members to be included in groups of hourly workers, or this particular issue may warrant no special attention at all. How much the union is involved in the Assessment effort depends on a number of factors, such as union-management relations, whether safety is an important issue for the union, and how crucial union support is for the Implementation effort.

Interviews at lower levels are done in groups of 8 to 12 people. This group setting is a comfortable one for most hourly workers, but they are often less accustomed to interview situations and may be reluctant to speak up in individual sessions. On the other hand, higher level managers may be more reluctant to talk frankly in a meeting of peers. There is usually also a fair amount of department-specific information which must be obtained from higher level managers, and a group setting is inefficient for this. Interviews usually last from 1½ to 2 hours, depending on the number of people, how talkative they are, and how interested they are in safety questions. Later interview sessions tend to be shorter as the interviewer finds that fewer new topics come up for detailed discussion.

Discovering the Basis of Perceptions

One of the challenges in interviews is to discover the basis of the perceptions that are revealed by the interviewees. Merely recording their expressed perceptions is not sufficient. It is important to inquire further and get specific examples when people make general statements. If a worker says, "it takes a serious accident to get any action on safety around here," the interviewer must find out what he or she is thinking about. This opinion might be based on a single incident that happened years ago, under previous management, with the situation long since corrected. Or the worker might be referring to a number of recent events which support the general statement.

In group interviews one of the concerns is to find out how representative the opinions are. A few outspoken people can easily dominate the discussion. Assertions made by only a few people need to be checked out with the other people present to see if they agree with them. For instance, a long-time worker might feel strongly that training is not adequate (based on a recollection of training received years before), but recently hired workers might think that training is very thorough.

The interviewers take notes during the interviews, reviewing them briefly immediately afterward to make sure that they recorded everything and that they understand what they wrote down. Later the notes are scored for themes or common issues, and the points from various interviews are brought together and summarized under the headings of the relevant themes. Differences by employee level in the organization or by department, shift, etc. are noted also.

One factor that affects how candidly people talk is anonymity or confidentiality. In some facilities people may be hesitant to speak up, fearing some kind of retaliation for expressing unpopular views. They need to be reassured that while their opinions will be summarized and reported in the Assessment Report, their remarks will not be attributed to anyone individually. People are oftentimes more candid with an interviewer who is from outside of the organization, and therefore does not have any vested interest in existing programs and who can impartially explore any topic that comes up.

Specific Areas to Assess

Management systems and accountability. This area provides rich ground for behavioral safety Assessment. Since an effective safety process is one that affects worker behavior on a day-to-day basis, responsibility for safety has to be at a level which facilitates this daily influence. This means that safety must be a responsibility of line management, with accountability for effective safety processes. At the facility level, establishing this kind of safety accountability begins when the facility manager meets regularly with department heads to discuss safety and to identify areas for improvement. Department managers in turn have to hold lower level supervisors responsible for specific outcomes. All too commonly, top management at a facility is committed to safety, but that commitment is not shared in the form of specific actions at lower levels. This is a failure of the management system to provide the necessary antecedents (objectives) or consequences (follow-up and review of performance). Inadequate management systems may reflect cultural factors— the widespread belief that safety should be the responsibility of the safety department or the position that department managers should have complete latitude to do as they think best in running their departments.

Maintenance and repair. Rather than focusing on unsafe behaviors as the causes of injuries, workers tend to focus on conditions and equipment. Consequently, workers often judge management commitment to safety by how quickly equipment and facilities are repaired and serviced. The workers do not see the monthly safety meeting run by the site manager. What they see are the leaks, and the machinery without guards. The workers are regularly around equipment that might seem unsafe to them. For this reason, maintenance has symbolic importance to workers. If the maintenance function is perceived as being unresponsive, workers are likely to think that they work in a place where management does not really care about safety. This perception often translates into cutting corners. While the position is not particularly logical, workers often have the attitude, "if management doesn't care, why should I?" Assessing the prevalence of this attitude is an important task of the interviews.

It is important for the Assessment effort to look into a facility's procedures by which facility and equipment problems are identified, reported, addressed, and followed up. Related questions are: How are maintenance and repair priorities set? Does maintenance have adequate staffing? What kind of feedback on job status is there from maintenance to the production organization? Oftentimes repair work is deferred for good reasons, such as needing to order a replacement part. But does this information get back to the person who originated the work order? A common response of Maintenance personnel is, "Well, if the production organization wants to know about the status of a job, they can always ask." This may be true, but the point is that by not having a proactive feedback system on repair items, the management system creates an atmosphere of unresponsiveness and lack of concern. The result is often that not only do people not ask about the status of a work order, they become indifferent about initiating work orders in the first place.

Discipline, rewards, and feedback. Discipline, rewards, feedback—each of these is a type of consequence for behavior. The key Assessment question is, how systematically are these consequences used? It is not uncommon for discipline to be used as a consequence for serious safety violations that result in an injury but very little is used for behaviors that have not yet hurt anyone. For example, unsafe lifting may be completely ignored as a safety hazard, even though an accident involving unsafe lifting is sure to receive a great deal of attention. In facilities that have hazardous production processes where unsafe acts by one person can expose many others to injury, workers may complain about lack of disciplinary enforcement of safety rules.

In both of these cases—unsafe lifting and hazardous processes—it is important for the Assessment effort to find out why appropriate discipline is not used. The reasons may be quite various. They could have to do with a militant labor union which fights all disciplinary actions, including safety-related discipline. The reason might lie in the location of the facility—in a small town where hourly workers and foremen grew up together and live in the same neighborhoods. The reason could be lack of clear policies about discipline, unclear safety standards, or a perceived lack of support for discipline by upper management. Most typically, discipline is not used very much for safety-related behaviors, and no one complains about it.

Reward systems are often used. The most common reward system is one in which if the facility or department goes for a specified period of time without a lost time injury or a recordable injury, then everybody in the facility receives a jacket, a cap, a dinner, or something of that type. Do such reward programs have the desired effect on the workers? Or are these reward programs symbolic only? Managers tend to think that such reward programs have a beneficial effect on safety performance—workers rarely see it that way. A sign of this difference is that hourly workers, and foremen too, often do not remember the conditions under which rewards are delivered. They cannot name how long their department must go without an injury; they are not clear about what kind of injury counts. The workers' unfamiliarity with the details may mean several things. It could mean that the programs do not have clear rules, that the rules vary from department to department, that payoffs come so seldom that workers are out of contact with the program, or that the programs are simply ignored. On the other hand, such programs are often the source of complaint—notoriously unsafe workers get the same reward as conscientious workers, etc.

Feedback for safe performance. Feedback for performance is a powerful consequence that is almost universally neglected. Workers often say that they never hear about safety outside of meetings, except that a foreman might sometimes criticize them if they are doing something unsafe. Foremen may agree that they rarely say anything to individuals about safety-related behavior. However, the same foremen also complain that they seldom hear about safety unless someone on their crew gets injured. Groups of foremen in well managed, international companies have been known to laugh out loud when they were asked whether they received any praise or recognition for good safety performance.

These chains of influence may go right up to the top of the management structure. A facility manager who is threatened with unpleasant consequences unless safety improves is understandably inclined to pass the threats down the line. If the threats reach the foremen they are faced with a dilemma. On the one hand a foreman is simply not going to give a written reprimand, or worse, to a worker who is out of shape and who does not lift properly. And yet, back injuries might be a major problem at the facility. Faced with this situation the foreman might turn to other means of reinforcing safety, but in practice it is hard for a foreman to use other means of influence when everybody else, including the boss, is focused on punishment.

Finally, even if there is no focus on punishment, good safety practices may simply go unrecognized. Most managers operate on a negative-exception basis. If there is an injury, attention is given to safety. Otherwise, not much is said. The Assessment interviews must handle this topic.

Production pressure. The sternest test of a commitment to safety comes when correcting a safety problem means lost production. Workers attend more to what managers do than to what they say. If they think that production controls everything else, they may say such things as, "management talks about safety, but. . . ." Production pressure can affect safety through maintenance or through staffing levels, too. For instance, a plant producing at peak output levels may not have any forklifts in reserve. Suppose then that the brakes on one forklift become marginal faster than is normal or anticipated. To take the forklift out of service may mean that product does not get moved. The drivers may be told that the brakes are not really a problem, that they should continue to use the unrepaired machine. Lean staffing levels may mean that there is only barely enough time to follow safe procedures. For instance, production foremen may put inappropriate pressure on overworked mechanics to get equipment back on line.

The true values of an organization are highlighted by finding out such things as who can shut down a line for safety reasons. Do hourly employees or foremen really believe that they have the authority to do it? Has a line ever actually been shut down for safety reasons? What were the consequences to the person who shut down the line for safety reasons? A foreman may say, "Yes, I can shut something down if I think it is unsafe, but I would have to explain personally to the facility manager why I did so." Such a foreman is likely to think three or four times before taking action. Continuing to run an unsafe line would have an injury as an uncertain consequence, but shutting down the line would have the certain consequence of having to justify the action to the facility manager. In such a situation, many foremen play the odds.

Four Keys of Effective Interviewing for Assessment

- Stay focused.
- Listen actively.
- Take good notes.
- Be in charge.

Staying focused. The following set of questions focuses the Assessment on the core factors under review. The questions are open-ended to encourage people to think about the subject. This means that the answers sometimes lead far afield. Keeping the established time constraints in mind, the interviewer follows where the answers lead. In order to assure both comprehensiveness and comparability of interview results, the interviewer completes all of the questions for each group of interviewees.

Sample Interview Question Set:

- What has worked well to make the plant/facility/department safer?
- What improvements could be made for safety?
- What would stand in the way of such improvements?
- How do you tell whether your safety performance is good?
- What happens if there is an accident?
- What proportion of injuries are reported?
- What one thing could your superiors/subordinates do to most improve safety?
- What is the most important thing that you could do to improve safety?
- Who has the authority to shut down an unsafe operation?
- Can you refuse to do unsafe work without repercussions?
- Do you ever do unsafe acts? Why?
- What drives the safety effort here?
- Who talks to you about safety? Under what circumstances?
- How much of your time is devoted to safety?

Listening actively. The interviewer listens attentively to what is said, being careful not to jump to conclusions or to assume that he or she already understands the emerging issues which the participants are voicing. At this stage, the behavior-based safety process is just being adapted to the facility, and the interviewer's job is to understand the cultural forces that predispose people either to behave safely or unsafely and to identify facility strengths and weaknesses as they apply to the Implementation effort. Assumptions are unwarranted and so the accurate interviewer asks open-ended questions, listens for nuances in the answers, and probes further when he senses unclearness either in the answer or in his understanding of the answer.

Taking good notes. Good notes are essential to the interview process. A great deal of information is communicated during the interviews, far too much for the interviewer to keep straight and sort out by memory alone. The following note-taking format has proven very effective in conducting and summarizing these Assessment interviews. Figure 5-2 shows a sample note page with entries on issues which typically arise in the interview situation. Figure 5-3 shows the same page with the Themes extracted from the notes.

- Each page is divided into two columns—a Themes column 2-in. wide on the left, a Notes column taking up the rest of the page on the right.
- During the interview the interviewer uses the Notes column to record the points made by the participants in answer to the Question Set.
- After the interview session, the interviewer goes back through the notes circling key words and ideas.

Interview Notes

10/10/88 1 p.m.
Wage Level
Operations — 1

Themes	Notes
	They never get maintenance done fast enough unless someone gets hurt. Then it gets done right away. It's hypocritical for them to say that safety is #1.
	Most people are safe most of the time. There are some flakes. They shouldn't be tolerated but they are. We all know who they are and we avoid them.
	Turn–arounds are bad times because it gets so chaotic. If there was better planning, that would help.

Figure 5-2. Interview notes.

- The interviewer then groups these key words and ideas into Themes, giving each Theme a short name and entering the relevant names in the Theme column beside the notes which they characterize. (See Figure 5-2 and Figure 5-3.) The result in a catalogue of the interview in terms of its important topics and issues.
- As soon after the interview as possible, the interviewer writes a paragraph summarizing his or her impressions of the key points, hot issues, potential problems, etc., as indicated by the interviewee/s. The interviewer may also want to write a second paragraph summarizing impressions about potential resources, leaders, committee members, trainers, critical behaviors, Observers, and other people who show promise of helping in the Implementation of the safety process.

Being in charge. The interviewer remains in charge of the interview throughout. This means that the interviewer:

- Sees to it that the meeting is positive and productive.
- Keeps the meeting lively.
- Ensures that no one participant monopolizes the group meetings.
- Keeps the meeting on schedule, moving through the Question Set.
- Conducts the meeting so that it does not become a mere gripe session. Recognizing that complaints are an important source of information,

---

Interview Notes with Themes	10/10/88 1 p.m. Wage Level Operations — 1

Themes	Notes
Maintenance **Hypocrisy** **Planning**	They never get maintenance done fast enough unless someone gets hurt. Then it gets done right away. It's hypocritical for them to say that safety is #1.
Discipline **Peer Relations**	Most people are safe most of the time. There are some flakes. They shouldn't be tolerated but they are. We all know who they are and we avoid them.
Planning	Turn–arounds are bad times because it gets so chaotic. If there was better planning, that would help.

Figure 5-3. Interview notes with themes.

the interviewer notes each kind as it comes up and then moves the interview on to the next issue.

- Gives factual answers to any question. This includes simply admitting it when the interviewer does not know the answer to a question.

THE WRITTEN SAFETY SURVEY

Interviews are useful for finding out about perceptions, for assessing intensity of feeling, and for identifying key issues. However, unless a very high proportion of the workforce is interviewed, it is difficult to compare the interview findings across departments, shifts, or employee level. Interview data is also relatively hard to quantify. Surveys provide a check on interview findings. They also provide a way of comparing opinions across groups. For these reasons interviews are supplemented with a written safety survey. The survey results are then analyzed by organizational level and department so that similarities and differences in perceptions can be discovered. Not only can the interviewer's impressions be cross-checked in this way, but communications blocks, polarizations, and other problem areas may be identified. When written safety surveys are used, they are administered at the beginning of the interviews.

The survey is anonymous. The supervisory or hourly respondent submits his completed survey anonymously, designating only his facility location or department and organizational level, hourly, supervisory, middle management, etc.

The survey that the authors have developed for use in their Assessment work is made up of 29 statements about key safety-related variables. The survey measures the respondent's perceptions about nine factors which are relevant to the success of the behavior-based safety process. The survey records the respondent's perceptions of:

1. The incidence of drug or alcohol use on the job.
2. The respondent's own level of involvement in existing safety efforts.
3. The existing level of positive reinforcement for safety at the facility.
4. Consistency of standards and enforcement about safety.
5. Conflict between production and safety—if any.
6. Management support for safety.
7. The causal relation between unsafe behavior and accidents.
8. The adequacy of existing safety training.
9. How readily safety-related facilities and equipment issues are addressed.

Whatever survey is used, the Assessment team looks for the patterns in the tallied scores—patterns of similarity and difference across employee level and location within the facility. It is important to note which employee levels answered similarly and differently. Similar patterns of answers by two employee levels may indicate identification or good communication. For instance, do supervisors see things more like hourly workers or more like middle managers? Very different answers between levels may indicate the breakdown of communication. When only one group sees things very differently from the others, it is important to find out whether the group in question has a special insight into the issue that has not been either widely communicated or appreciated by other groups, or whether the group is blind to the issue or wants to deny its existence.

By eliciting answers to these and related questions, the Assessment effort arrives at an informed evaluation of the facility's resources for Implementation. In addition, techniques of behavioral analysis are brought to bear on a representative sample of the facility's accident reports. In this way, facility management gains a clear picture of the most fruitful areas in which to focus the Implementation effort. Chapter 6 takes up this subject.

Chapter 6

Behavioral Analysis of Accident Reports

Behavioral analysis of accident reports lays the foundation of the facility's inventory of critical safety-related behaviors. This preliminary analysis of accident reports during the Assessment effort identifies the behavioral categories and items that are most often associated with the facility's accidents and injuries. A behavioral category is a generic class of behaviors such as Body Placement, Tool Use, Proper Procedure, Housekeeping, Personal Protective Equipment, etc. A behavioral item is a specific behavior that falls within a behavioral category. Under Body Placement one might find *Standing in front of discharge,* or *Reaching into a roller to clear a jam,* etc., Figure 6-1. These same findings guide the Steering Committee during the Implementation phase while they develop the facility's inventory of critical behaviors and its observation Data Sheet. Behavioral analysis of accident reports

Generic Behavior	Specific Behavior	Accident Number
Body Placement	Standing in front of discharge	87 – 02
	Reaching into Roller to clear jam	86 – 37
	Standing on chair	87 – 03
Body Mechanics	Bent and twisted to move drum	87 – 09

Figure 6-1. Identifying the behavioral categories associated with injuries.

gathers the information that the Steering Committee uses to decide such questions as:

- Should the facility's inventory of critical behaviors be developed by department or for the plant as a whole?
- Should the inventory have many different items on it or only a few?
- Should the items on the inventory be primarily job-specific, or should they be primarily generic items?

The formation of this Assessment group varies with each facility. Sometimes this group is the same as the informal group mentioned in Chapter 4—the group that initially sponsors the Assessment. In a facility where the Steering Committee begins to take shape before the Assessment effort addresses the task of behavioral analysis of accident reports, the group analyzing the facility's accident reports may be a subset of the Steering Committee. In either case, the group that does this work needs to be representative and small—5 to 10 people. The group needs representatives from each major area or department. These people should be very familiar with the work in their departments and have some knowledge of the accidents which have occurred there. Other important characteristics of this Assessment group are that:

- They express themselves well verbally.
- They think that safety is important.
- They are members or potential members of the Steering Committee.

A secondary aim of the behavioral analysis of accident reports is to provide the assessment group with training in the techniques of inventory development, techniques that come into extensive use during the Implementation effort.

FOCUSING ON ACCIDENT REPORTS

In 80%–95% of injuries, some type of unsafe behavior was the final common pathway of the incident. To reduce injuries then it is necessary to know what kinds of unsafe behaviors are occurring. Most companies collect various types of summary information on injuries, including information on injury type. Usually these injury types are named after the categories used for government reporting purposes, categories such as *struck by, struck against,* and so forth. However useful these categories are for government record keeping, they are not very helpful for injury prevention.

For instance, suppose that while tightening a bolt a worker suffered contusions, striking a hand against something when the wrench slipped. This is not a serious injury perhaps, but if the contusions were severe enough the worker might have difficulty holding hand tools for a while, restricting his or her work assignment until the hand healed. In this case, the accident report would probably classify this injury as an instance of *struck against*. However, what happened that caused the wrench to slip so that the worker struck and injured his hand? The possibilities are various, and most of them are of interest to the Assessment group. Here are three possibilities.

1. The bolt might have been rusted in place, or the head of the bolt might have been rounded off—either case calling for the worker to burn the bolt off with a cutting torch instead of trying to use a wrench on it.
2. Or the worker might not have seated the wrench properly on the bolt. Or perhaps he seated the wrench properly but then jerked on the wrench or applied too much force.
3. The worker might have been able to position the wrench differently so that in the event that it did slip there would have been no injury.

These three behavioral causes suggest such categories as: 1. Procedure, 2. Tool Use, and 3. Body Use.

It is far more useful to know in how many cases using an *improper tool or procedure* was the final common pathway of injury than it is to know how many injuries were caused when workers *struck against* something. The reason for this is that by the time a worker has struck against something, the damage is usually done. Supervisors who tried to use this *struck against* information proactively would find themselves using very vague terms and warning workers to be careful not to strike against things. The warning is not specific enough to be useful to the workers. It is like saying, *pay attention*. Furthermore, the supervisors would find it difficult to know whether the workforce was being careful not to strike against things. So the *struck against* category is not specific enough to be useful to the supervisors, either. The behavioral category, on the other hand, addresses all three of these points. Improper tool use or procedure is something that both the worker and the supervisor can be aware of before an injury takes place and the damage is done. The guidelines for proper tool use and procedure can be spelled out in sufficient detail that both the worker and the supervisor know what counts as safe behavior, and what counts as unsafe behavior. The same is true for proper body use, including body placement and position in respect to task. The aim of behavioral analysis of accident reports is to arrive at the set of behavioral categories that accounts for a significant portion of a facility's accidents. In most cases, even long-time employees at a facility are surprised to discover just which categories of behavior lie at the core of their recurring safety problem-areas. They are also usually surprised to discover how great a proportion of their total accident frequency is associated with this core of behaviors. This knowledge—and its importance to management efforts—is the incentive for Assessing the facility's accident reports carefully. It is important that the Assessment group bear this in mind because the behavioral analysis of accident reports can be tedious at first.

Information concerning the behavioral components of injuries is often lacking in standard accident investigation reports (a defect that is corrected as the behavior-based safety process achieves maturity in integrated accident investigations and safety meetings). When such information is present in standard investigation reports, it is rarely summarized. However, behavioral information can be extracted from accident reports fairly reliably. The procedure involves collecting a set of reports for a recent period of time.

Once the reports have been assembled, a group of 5 to 10 people analyzes the injuries. These should be hourly employees and others who are familiar with the tasks involved in the injuries. At a given facility, this requirement might mean that injuries are best analyzed by department or area. For each injury, the group does two things:

1. The group analyzes in behavioral terms what the worker was doing when he was injured; and
2. The group states what the worker might have done differently to avoid the injury.

Once the group catches its stride, it arrives at these two things for most of the accident reports. In only a small percentage of cases will the group fail to find factor 1 and/or 2.

The second part of the group's task results in a set of statements which specify what workers need to do to avoid injury. These statements are of the form, *When performing such-and-such a task, here is what the worker needs to do.* For instance, *When uncoupling hoses, first check that valves are closed and that product has been cleared from the hose.* Even at this preliminary stage of developing the facility's inventory of critical behaviors, the emphasis is on producing clear operational definitions of safe and unsafe behavior. Sometimes it can seem faster to focus solely on task #1—identifying in behavioral terms the causes of injuries at the facility. This is shortsighted because the final aim of the behavior-based safety process is to reduce injuries at a particular facility by reducing the number of unsafe behaviors performed there. To accomplish this aim, workers need to know the safe way of performing the important jobs at the facility.

Generalizing from specific circumstances.   To be most useful, the circumstances of particular injuries need to be generalized so that they apply to related situations. For instance, in response to three different injuries the Assessment group might produce the following three statements:

- When working on a ladder, it should be tied off.
- When working in a man-lift, wear a safety harness.
- When working on a catwalk, do not lean over the railing.

It is important for the Assessment group to generalize these statements, producing a statement such as:

- When working at an elevation, the working platform should be steady, the worker should be secured, and the work site should be immediately adjacent to the worker. (For instance, tie off a ladder or have a co-worker hold it; wear a safety harness in a man-lift; do not lean over the railing of a catwalk or over the edge of a ladder.)

As the assembled accident reports are analyzed, fewer and fewer new behavioral items have to be added. The behavioral components of injuries begin to sound very similar to the ones that have already emerged in the analysis of the other reports. Usually a fairly small number of items accounts for a large proportion of injuries. The Assessment Report, Chapter 8, communicates the results of the behavioral analysis of the facility's accident reports.

In addition to behavioral analysis of accident reports, the Assessment effort also employs statistical analysis of injury data to discover which of the factors of the facility's accident frequency are due to common cause and which are due to special cause. Chapter 7 gives a brief presentation of this important question.

Chapter 7

Statistical Analysis of Injury Data

There are three ways that statistical analysis of injury data can contribute to the Assessment effort.

Facts, not guesses. Statistical analysis helps to develop an accurate picture of precisely which variables are predictors of accident frequency. This knowledge separates fact from myth and guesswork.

Focus. Statistical analysis of injury data focuses attention on those facets of the facility's accident picture that are most significant.

To establish a baseline. Statistical analysis of injury data allows management to establish a baseline against which they track their safety efforts during Implementation and beyond, documenting success and indicating where adjustments are needed.

Many people understand these advantages of statistical analysis of injury data but nonetheless fail to approach the task systematically. They behave as though they regard their impressions as sufficient. Reliance on impressions is counterproductive, though, because impressions can be imprecise, even blind, about opportunities for improving the facility's injury frequency. This is not really surprising since the number of variables and their combinations are too complicated to keep track of except by computer. Software for statistical analysis of injury data is available from a number of sources, including the authors. Even with the added power and speed of the computer, however, the ultimate value of the analyses depends on the quality of the information that is entered into the system. Information of high quality is needed. The following are some the principles of high-quality injury data information.

Five Principles of Injury Data Analysis

Of course there are many principles of statistical analysis. The following five, however, represent a minimum set for working proactively with injury data:

- Define the terms.
- Look for the rate.
- Check the accident frequency rate for all relevant variables.

- Cross atypical variables with standard variables.
- Use statistical techniques to determine significance.

1. Define the terms. The first principle of statistical analysis of injury data is to define one's terms and parameters. Otherwise vagueness and misinformation are the result. For instance, suppose that someone reports that, *95% of this* facility's *accident frequency is accounted for by employees with multiple incidents, and these employees constitute only 20% of the workforce.* What does this tell us? Very little unless one knows what the speaker means by *accident* and within what period of time an incident counts as a *multiple incident.* Do first aids count as accidents? Does the statement cover a 2-year span (a helpful parameter for identifying employees who have a pattern of multiple incidents), or a 10-year span (too long a period of time for identifying multiple incidents unless a high number of incidents per worker is used).

2. Look for the rate. Suppose a safety publication prints the statement that at XYZ Company 47% of the recordable injuries are due to truck drivers. This statement does not tell nearly enough to be a trustworthy indication of where to focus an Implementation effort to have the most effective impact on safety performance at XYZ Company. By itself the 47% figure indicates almost nothing of importance. At the least the statement must also say what proportion of the workforce at XYZ is made up of truck drivers. Do they represent 47% of the workforce, 94%, or 23.5%? But even beyond these numbers, it is necessary to know how many of the total hours worked at XYZ Company are worked by truck drivers. This illustrates the second principle of statistical data analysis—counting injuries is not enough, the injury frequency *rate* is needed. Looking at frequency rates means taking the exposure hours into consideration. The accident frequency rate includes exposure hours in its definition—accident frequency rate equals the number of incidents per per 100 employees per year (or per 200,000 man hours worked).

Although most companies realize that the incident count is only part of the accident frequency rate, they do not carry through with this realization. Having arrived at the accident frequency rate for an entire facility they stop just when they should continue with a more detailed statistical analysis of their injury data. Compared to the potential value of analysis of accident frequency rate by variable, the rate for an entire facility is a very crude measure that is of limited use. This brings up the third principle.

3. Check the accident frequency rate for all relevant variables. The best way to use frequency rate analysis is to examine the frequency rates of variables, such as Age, Years-on-the-job, Shift, Gender, Occupation, Time-of-day, Day-of-rotation, etc. Rate is a straightforward calculation for any variable that has exposure hours associated with it. In addition, there are atypical variables that do not have exposure hours, for example, Part-of-Body. There are no more exposure hours for hands than for feet. The way to analyze the rate for atypical variables is to cross them with standard variables.

4. Cross atypical variables with standard variables. By crossing an atypical, non-exposure-hour-variable such as Part-of-Body with a standard variable, such as Department or Shift, an accident frequency rate can be analyzed for the atypical variable in relation to the standard variable. In this case the rate would be for Body-part injuries by Department or Shift.

Having calculated the Part-of-Body by Department or Shift, it is possible to analyze for differences. For example, hand injuries may be three times as frequent in Department A as in Department B. This difference between departments seems significant at first glance, but at this point it is not really known whether it is. The question now is, How does one know whether the findings are significant or whether they are due merely to random fluctuations?

5. Use statistical techniques to determine significance. Statistical techniques for distinguishing significant from random variation are available. There are mathematical methods of determining whether the apparent findings are random differences or whether they reflect real underlying causes. These methods are relatively easy to apply using statistical computer software designed for this purpose.

Organizing Accident Data

Using a computer to do accident data analysis can solve many problems, but it requires that the data be well organized before it is put into the computer. The accuracy of computer reports can be no better than the accuracy of the data that goes into them—*garbage in, garbage out.* Data must be clean. Clean data is:

Organized. Data should be summarized on a log or listing.

Complete. All relevant information is included.

Consistent and Unambiguous. Due care is exercised about information such as names and ages. Names are easily misspelled or confused. Two people may have the same last name and the same first initial. Uniquely assigned numbers are better—social security numbers, for instance. Ages, such as thirty-two years old, are vague: Date of Birth is better.

Legible. Data entry personnel cannot enter what they cannot read.

Accurate. Estimates are allowable within limits, but data gatherers need to make every effort to collect the most accurate figures possible.

Exposure Hours and Accident Data

Two categories of data are important to the Assessment effort, Accident Data and Exposure Hours. Both need to be supplied in a log format (see Figure 7-1 and Figure 7-2). The accident log supplies the following information.

ACCIDENT DATA LOG
(Sample)

	Emp#	Name	DOB	DOH	Cost	DOI	Rec	LD	RD	CB's	1	2	3	4	5	6...
1.	565-56-2886	JONES,T	042041	090185	2000	080285	Y	300	50	1.13	01	05	13	01	07	10
2.																
3.																
4.																
5.																
6.																
7.																
8.																
9.																
10.																
11.																
12.																
13.																
14.																
15.																

Figure 7-1. Accident data log.

A unique identifier. Each employee who was injured is listed with a unique identifier, such as their social security number.

Dates. Use the MMDDYY format for any variable involving a date—birth date, date of hire, date of injury.

Severity. In order to generate severity reports, injuries need to be coded as either Recordable or First Aid. Number of days lost and/or restricted need to be supplied along with other identifiable costs of the injury.

Mutually exclusive, exhaustive elements. The elements of a variable should be mutually exclusive and exhaustive of all the possibilities for the variable. With variable elements that are not mutually exclusive, some exposure hours are counted twice, giving an erroneous frequency rate. With variable elements that are not exhaustive, some relevant exposure hours may not be counted at all, also giving an erroneous frequency rate.

The Exposure Hours Log

The exposure log is a listing by variable element with associated hours worked for each element. Provide the start and end dates covered by the hours. These dates must exactly match the periods for which accident data is supplied. The variable and item names in the Exposure Hours Log need to be

EXPOSURE HOURS DATA LOG
(Sample)

Variable Number	Variable Name
5	MONTH of YEAR

Element Number	Element Name	Exposure Hours
1	January	84,940
2	February	76,720
3	March	84,940
4	April	82,200
5	May	84,940
6	June	82,200
7	July	84,940
8	August	84,940
9	September	82,200
10	October	84,940
11	November	82,200
12	December	84,940

Figure 7-2. Exposure hours data log.

identical with the variable and items names in the Accident Log. Estimating exposure hours is sometimes necessary, but it must be done as accurately as possible since the value of the results is limited by the accuracy of the input.

All of these issues are implied in the structure and procedures associated with data logs. Assessment teams who are unfamiliar with these procedures do well to secure the advice and help of someone who is familiar with statistical analysis. It is much easier to correct a misunderstanding early than to make up for it later. For instance, note the variable number and name on the Exposure Hours Data Log in Figure 7-2. Each variable of interest for analysis needs to have its own Exposure Hours Log page/s.

Analysis of Data

Compiling the information data as described above allows a statistical software package to produce accident frequency analyses as reports. These reports need to include statistical testing to determine whether observed differences in rates are *significant* statistically or whether they occur by chance. In SPC terms this distinction is referred to as "special cause" versus "common cause." To say that a difference in accident frequency rates is due to common cause is to say that the variation is basically random. Variation due to special cause, however, is not a function of some random fluctuation within the system but the result of something outside the system.

For example, suppose management notices that accident frequency rates in July and August appear to be consistently higher than during other months. Is this the result of something that happens during July and August (special cause), causing the frequency rate to rise—the weather, more turnarounds, etc? Or is the variation just a chance fluctuation in the frequency rate? This is a crucial question for management to answer, because their actions will be guided accordingly. If management acts when the fluctuation is random only, they create myths about accident frequency, solving "problems" that are not even there. This kind of activity is misleading and counterproductive. On the other hand, when in fact a special cause does exist and management fails to act, they miss an opportunity to reduce accident frequency.

Common cause or special cause—how to tell them apart? Statistical testing is the method used to answer this question. Over-simplifying the subject a bit, one could say that in quality improvement efforts the most common procedure is a control chart in which variability within "upper and lower limits" are taken as common cause, and anything outside the limits is understood as special cause.

An appropriate method of statistical testing for accident frequency is Chi-square analysis. (See any basic statistics text for an explanation of the Chi-square statistic and its use.) A complete explanation of Chi-square analysis is beyond the scope of this book. The basic idea can be simply presented, however. In Chi-square analysis the likelihood that a common cause exists is stated as the probability that the observed variation is "significant." It is a convention that any variation over a probability of 0.05 casts enough doubt on itself to be considered common cause. In other words, only variations whose probability is 0.05 or below are certain enough to warrant taking them as due to special cause.

What this means in practice is that if the statistician, or the computer report, says, "These numbers are significant at the 0.05 level (or below)," "management can proceed in the confidence that they have indeed identified a real problem (see examples below). If the statistician or computer report says, "These numbers are not statistically significant," then management should withhold action on the merely apparent special cause and instead focus on improving the common safety system that is generating the accidents.

Figure 7-3 through Figure 7-9 show a set of reports prepared using Chi-square analysis and showing the statistical significance level for each report. These are actual reports produced by the authors for a chemical company. Figure 7-3 shows data for the variable "Year," comparing accident frequency for 1986, 1987, and 1988. The report indicates the changes from year to year are not statistically significant. In other words, they may well be random variation. The manager who thought that the 1988 rate of 2.87 represented a significant improvement over 1987's rate of 3.81, would risk being incorrect.

Frequency, Plantwide by Year

1 / 1 / 86 — 12 / 31 / 88   page 1

Hours Worked	Recordables	Year	Frequency Rate 0—5—10—15—20—25 :—:—:—:—:—:
2778255	34	1986	2.45]]
2308761	44	1987	3.81]]]]
1113554	16	1988	2.87]]]
6200570	94	Totals	3.03

Chi Square: p < .20 Not Significant

Figure 7-3. Frequency—facility-wide by year.

		Frequency, Month		
1 / 1 / 86 — 12 / 31 / 88				page 2
Hours Worked	**Recordables**	**Month**	**Frequency Rate**	0----5---10---15---20---25
				:----:----:----:----:----:
638787	5	Jan	1.57	]]
580157	8	Feb	2.76	]]]
644111	13	Mar	4.04	]]]]
581306	13	Apr	4.47	]]]]
620023	5	May	1.61	]]
590323	8	June	2.71	]]]
420485	**16**	**July**	**7.61**	]]]]]]]]
442991	5	Aug	2.26	]]
422907	7	Sept	3.31	]]]
429577	3	Oct	1.40	]
419829	6	Nov	2.86	]]]
410074	5	Dec	2.44	]]
6200570	94	Totals	3.03	

Chi Square: $p < .01$ Highly Significant

Figure 7-4. Frequency—by month.

In fact, the manager's system was stable and would predictably generate about three recordable injuries per 100 employees per year. This would continue indefinitely until the system itself was improved.

Figure 7-4 shows accident frequency by month of the year, collapsing over three years. Note that the month of July stands out and is significant statistically.

Frequency, Shift				
1 / 1 / 86 − 12 / 31 / 88				page 3
Hours Worked	Recordables	Shift	Frequency Rate	0—5—10—15—20—25
				:—:—:—:—:—:
966408	19	A	3.93	]]]]
966408	20	B	4.14	]]]]
966408	12	C	2.48	]]]
966408	18	D	3.73	]]]]
2344935	25	Days	2.13	]]
6210567	94	Totals	3.03	
Chi Square: p < .20 Not Significant				

Figure 7-5. Frequency—by shift.

Therefore the plant manager could proceed with confidence to focus on that time of year and find the "special cause" for corrective action. In the established behavioral safety process one way of carrying out corrective action would be to increase behavioral observation frequency during July.

Figure 7-5 shows that Shift is not a factor and, therefore, does not warrant increased attention. Figure 7-6 shows that Time-of-Day is a significant factor and does warrant special attention. Again, increased observation frequency would be a good strategy. Figure 7-7 and Figure 7-8 show that neither Age nor Gender is a factor that reliably predicts accident frequency at this plant.

Figure 7-9 shows that Department is a significant variable, and this warrants further attention. Perhaps behavioral analysis is needed for relevant management system issues or specific jobs in the high-risk departments.

Additional analysis would focus on Part-of-Body, accident type, or some other variable, in connection with the variables shown in Figure 7-3 through Figure 7-9. For example, in the case of the company whose data is shown here, the authors found that certain departments had elevated accident frequency due to hand injuries, while other departments had elevated frequency due to back injuries. This information helps to shape the facility's

Frequency, Time of Day

1 / 1 / 86 — 12 / 31 / 88   page 4

Hours Worked	Recordables	Time of Day	Frequency Rate 0----5---10---15---20---25 :----:----:----:----:----:
486460	6	8:00 am	2.47]]
486460	4	9:00	1.64]]]
486460	8	10:00	3.29]]]
486460	5	11:00	2.06]]
486460	9	12:00 noon	3.70]]]]
486460	9	1:00 pm	3.70]]]]
486460	4	2:00	1.64]]
486460	3	3:00	1.23]
486460	1	4:00	0.41
143260	8	6:00	11.17]]]]]]]]]]]]
143260	7	7:00	9.77]]]]]]]]]]
143260	5	8:00	6.98]]]]]]]
143260	4	9:00	5.58]]]]]]
143260	2	10:00	2.79]]]
143260	2	11:00	2.79]]]
143260	1	12:00 midnight	1.40]
143260	2	1:00 am	2.79]]]
143260	4	2:00	5.58]]]]]]
143260	1	3:00	1.40]
143260	1	4:00	1.40]
143260	2	5:00	2.79]]]
143260	2	6:00	2.79]]]
	4	*unknown*	
6240520	94	Totals	2.88

Chi Square: p < .01 Highly Significant

Figure 7-6. Frequency—by time of day.

inventory of critical behaviors and also tells the Observer which behaviors to target during observation.

Where the target population is large enough, multi-variate analysis can be performed. This subject goes beyond the scope of this chapter, but it is mentioned as an approach that offers important potential benefits. In one metropolitan transit company, for example, the authors found that female

Frequency, Age

1 / 1 / 86 – 12 / 31 / 88 page 5

Hours Worked	Recordables	Age	Frequency Rate	0—5—10—15—20—25 :—:—:—:—:—:
240405	3	31 – 35	2.50	]]]
1082814	18	36 – 40	3.32	]]]
930943	12	41 – 43	2.58	]]]
1395234	20	44 – 47	2.87	]]]
1057894	13	48 – 51	3.97	]]]]
733683	5	52 – 55	3.54	]]]]
409178	2	56 – 59	2.44	]]
6213573	94	Totals	3.03	

Chi Square: $p < .20$ Not Significant

Figure 7-7. Frequency—by employee age.

Frequency, Gender

1 / 1 / 86 – 12 / 31 / 88 page 6

Hours Worked	Recordables	Gender	Frequency Rate	0—5—10—15—20—25 :—:—:—:—:—:
3859430	54	Male	2.80	]]]
2351140	40	Female	3.40	]]]
6210570	94	Totals	3.03	

Chi Square: $p < .20$ Not Significant

Figure 7-8. Frequency—by gender.

Frequency, Department

1 / 1 / 86 — 12 / 31 / 88   page 7

Hours Worked	Recordables	Dept.	Frequency Rate
			0----5---10---15---20---25
			:----:----:----:----:----:
785008	23	Spinning	5.86]]]]]]
643895	20	Drawjet	6.21]]]]]]
1073527	16	Engineering	2.98]]]
350744	1	Personnel	0.57]
212212	10	HSJT	9.42]]]]]]]]]
388893	4	Inspection	2.06]]
643977	7	Drawtwist	2.17]]
77355	4	Day Crew	10.34]]]]]]]]]]
145266	1	Polymer	1.38]
78968	2	Denier Room	5.07]]]]]
421101	1	Block Room	0.47
86169	1	Pack Room	2.32]]
387998	1	Qual. Control	0.52]
125046	2	Type G	3.20]]]
	1	*unknown*	
6100031	94	Totals	3.03

Chi Square: p < .01 Highly Significant

Figure 7-9. Frequency—by department.

bus drivers had significantly higher accident frequency than male bus drivers during their first three years on the job and significantly lower accident frequency than male drivers after three years on the job.

Ultimately a prediction model can be developed which gives weight to each variable in accordance with its ability to predict accident frequency. In this way each worker's risk can be assessed, in much the same way that the health risk of heart attack is currently assessed based on predictors such as weight, diet, fitness, family history, stress, etc. In one prediction model, the authors found that Age, Occupation, and Department all interacted with each other in predicting accident frequency. For instance, as a category of Occupation welders were only moderately at risk compared with other categories of Occupation. But welders in one Department had a risk factor elevated by a factor of 4, and welders under Age 28 had a factor elevated by a factor of 3.

SUMMARY

Techniques of statistical analysis, some of them quite sophisticated, have been used in industry for many years now, and they might have been applied to injury data long ago. However, in the absence of management techniques that could make use of the statistical analyses, there was no demand for such analysis. Even among companies who are familiar with the SPC approach to quality improvement, there has been very little initiative to apply these principles to safety performance. The development of the behavior-based approach to safety management changes this status quo.

Chapter 8

The Assessment Report

The Assessment effort produces an Assessment Report. The Assessment Report serves several functions. Most obviously it conveys the findings from the safety interviews and surveys, and from behavioral analysis of accident reports and statistical analysis of injury data. Reporting formats for these findings are presented below. First, however, a word about the planning function of the Assessment Report. The effective Assessment Report is structured as an Implementation Plan, culminating in a Timeline of suggested steps forward. The information gathered during the Assessment effort is organized and reported as part of an overall proposal about how the particular facility should proceed with implementation of the behavior-based continuous improvement process. The challenge for the Assessment group and the Steering Committee therefore is to adapt the characteristic steps of Implementation and the four basic functions of the behavior-based process to their particular facility with its organization and needs. Figure 8-1 represents the elements that must be coordinated during the Implementation effort. From the site manager to hourly employees, the roles and responsibilities of the organization are treated in Chapter 9. Chapters 10-16 are concerned with the steps of Implementation and with the four functions of the behavior-based continuous improvement process. Chapter 17 provides five Case Histories of Implementation efforts.

The primary purpose of the Assessment Report itself is to inform and engage the support of the facility's managers and Steering Committee. While it is effective, and a good sign even, for the Assessment effort to be sponsored by an informal group of key facility personnel, the Implementation effort itself needs the sponsorship of the site management team. The Assessment group needs to be aware of this, and therefore to formulate an Assessment Report that addresses the known concerns of the facility management team.

The cogent presentation of the Assessment findings will go a long way toward meeting the concerns of any management team that sees improvement in safety performance as an area that offers the facility an opportunity for significant gains.

Reporting the Interview and Survey Findings

The readership of the Assessment Report needs to know in general terms both the interview and survey procedure, and the results. Procedural matters

Roles and Responsibilities	5 Steps of Implementation	4 Functions of the Safety Process
Site Manager	1. Develop the facility inventory of critical behaviors	• Planning
Middle Managers		• Observing
First-line Supervisors	2. Conduct Inventory Review Meetings	• Facilitating Meetings
Hourly Employees		• Training
	3. Train Observers	
	4. Conduct Kickoff Meetings	
	5. Establish the continuous improvement process	

Figure 8-1. The elements that require coordination during Implementation.

are such things as the numbers of employees interviewed from which levels and locations within the facility and might include a list of open-ended questions which the interviewees were asked to respond to. The Assessment Report stresses the fact that the interviewees represent a cross section of the facility and that the interviews and surveys themselves were conducted so as to elicit and accurately record the perceptions of that cross section.

The themes which emerged from the interviews can be effectively organized under such headings as *Plant Strengths* and *Plant Challenges and Concerns*. Each theme represents a frequently voiced perception. When reporting the results of the interviews and the surveys, the emphasis is that although there may be inaccuracies in employee perceptions, it is nonetheless true that the themes of the interviews are clearly things that are important issues in people's minds. As such they are issues that will need to be attended to during the Implementation effort. This is not, however, to say that Assessment uncovers only difficulties and weaknesses. An effective safety Assessment effort finds what is present in the situation of the facility, and most facilities have quite a diverse fund of strengths and of safety capital, so to speak.

Plant strengths. The behavior-based safety Assessment effort is geared to discover the forces and factors present in the facility which favor the Implementation of the continuous improvement safety process. Following this particular line of sight, the Assessment often turns up strengths that facilities did not know they possessed. The effective Assessment Report presents these areas of strength clearly and with an eye to the way that they will play into the Implementation effort. For instance, the behavioral safety Assessment of a highly functioning facility might report plant strengths in areas such as *Attitudes toward behavior, Communication and cooperation, Continuous improvement process philosophy, Power to stop production,* and *Safety commitment.* An *attitude toward behavior* that is a plant strength is one that understands that behavior has a great deal to do with safety. In facilities where this is the prevailing attitude of the safety culture, efforts to manage safety performance proactively will probably succeed. *Communication and cooperation* as a plant strength refers to the quality of the relations between levels and among departments of the facility. Since Implementation of the behavior-based process requires organizational development, it places a high premium on open communication within the organization. As its name indicates, behavior-based safety management is a process approach to safety, and this means that a facility that values the *continuous improvement process* as a mode of operation—perhaps in quality improvement efforts, perhaps in other areas—is in a stronger position for Implementation. When the *power to stop production* inheres in the work group, it represents a plant strength because such a work group is partly self-regulating, and the mechanism of continuous improvement in safety is a function of a self-regulating work group that is focused on its own safety performance. *Safety commitment* might mean a commitment to the possibility of achieving an injury-free environment, an obvious plant strength when it comes to Implementation of the safety process.

Plant Challenges and Concerns. The Assessment Report also must spell out the issues and concerns that emerged from the interviews and surveys. It is not just a slogan to refer to these matters as challenges rather than as weaknesses or problems. Insofar as these interview and survey findings represent facility conditions and factors that hinder or retard Implementation, they are problem areas. However, identifying them is already helpful since it is the first step in working with them and turning them to good account. A behavioral safety Assessment of a facility might report plant challenges and concerns under such headings as *Production pressures, Safety is adversarial, Inconsistency of standards and practices,* and *Management and safety department visibility. Production pressures* are a challenge in many facilities. The real test of commitment to safety is usually the production schedule. The old notion that production and safety are incompatible is a liability both for safety and for production and so this kind of production pressure on safety represents a challenge that must be addressed if Implementation is to succeed. The perception that *Safety is adversarial* is an important theme because it indicates a safety culture in which blame and fault-finding are too prominent in people's minds. Whether or not safety really is adversarial at the facility, if many people believe that it is, it may as well be for all practical purposes. This is a challenge that the behavior-based approach is well suited to handle because of its emphasis on anonymous, random sampling of the work place. *Inconsistency of standards and practices* is a challenge because a workforce that is sensitive to the presence of double standards in safety— between levels and/or across locations—needs to be assured that the facility inventory of critical behaviors is going to be applied objectively and consistently. *Management and safety department visibility* is a challenge for any facility that embarks on the behavioral safety process. The perception may be that management does not really care about safety and that the safety department only cares about programs or paperwork. It is very important for successful Implementation that the workforce not see the continuous improvement process as yet another program, here today and gone tomorrow. The remedy is genuine management involvement in safety.

Reporting Behavioral Analysis of Accident Reports

The readership of the Assessment Report needs to know the general procedure that was used in the course of the behavioral analysis of accident reports. This includes such matters as indicating who served on the Assessment group, how many reports were analyzed, and which years were covered by the accident reports. Also indicated are how many behavioral items and categories were extracted from the accident reports, and of these items and categories how many were worked up into operational definitions for the facility inventory of critical safety-related behaviors.

Findings. Under findings it can be very helpful for plant management and the Steering Committee to see which behavioral items and categories were associated with the largest proportion of the facility's injuries. Typically, a cluster of 4-7 behavioral categories accounts for a significant majority of a facility's recurrent safety problem areas. The Assessment Report stresses the fact that by focusing the continuous improvement process on these several behavioral categories, the facility is assured of achieving demonstrable success in its safety effort.

Reporting the Statistical Analysis of Injury Data

Procedure. This section of the Assessment Report needs to detail the procedure followed to gather and analyze the facility's injury data.

Findings. The purpose of statistical analysis of injury data is to look for especially promising areas on which to focus. For instance, statistical analysis of injury data helps to identify the times and places when it is most effective to increase the frequency of behavioral observations. Typical findings are ones that show statistically significant variations in injury frequency as related to the variables that were analyzed, e.g., Time-of-day, Day-of-week, Overtime, Experience level of employee, etc.

Recommendations and Suggested Steps Forward

Recommendations. On the basis of the findings from interviews and surveys, and from behavioral analysis of accident report and injury data, the Assessment Report makes recommendations about the scope and administration of the Implementation effort. It must also address such topics as Training Requirements, Steering Committee selection, Inventory development, Data management, Time requirements, and how these matters relate to the cultural issues uncovered by the Assessment.

Steps Forward. The Assessment Report may include a presentation of suggested steps forward. The following is an example.

1. Assessment review and strategy session
2. Selection of Steering Committee, Committee Facilitator, Trainers, and Observers
3. Training for Observers, Managers, Supervisors, and Trainers
4. Development of the behavioral inventory
5. Inventory Review meetings
6. Observers gather baselines
7. Steering Committee prepares for Kickoff Meetings
8. Kickoff Meetings are held for all employees
9. Steering Committee evaluates Implementation effort, makes recommendations for establishing the ongoing safety mechanism

Chapter 9

Roles and Responsibilities

Successful Implementation of the behavior-based safety process at a particular facility involves the clear handling of roles and responsibilities of two kinds—1) organizational responsibilities to the Implementation effort, and 2) the roles and responsibilities of the safety process itself.

The organizations that are most amenable to behavior-based safety management are the ones where responsibility for safety already clearly resides with line management, and where there is an ongoing management system that includes accountability for safety-related issues. In these cases implementing the behavior-based approach is easier because the specific goals of Implementation can be included as part of the already established goal-setting and accountability structure.

The facility manager ultimately has responsibility for the success of Implementation. Working with department heads the manager sets the specific criteria for support of Implementation. These departmental criteria in turn imply supportive roles from lower level managers down to the level of first-line supervisor. Effective management avoids the common problem with safety goals—setting them in terms of the wrong objectives. Instead of setting goals such as specific %Safe ratings or specific injury rates, proper early objectives are activities and outcomes which support Implementation itself. For example, response to safety-related maintenance items identified by Observers. In facilities that lack a working, visible system to respond to safety-related maintenance items, when Observers first start using the facility Data Sheet they oftentimes identify a number of items as requiring correction, repair, etc. For the continuous improvement process to be credible, these items need good follow-up from management. Where such follow-up is lacking, workers typically conclude that the focus on behavior is just a management substitute for spending money fixing equipment. This conclusion is common even where most injuries are unrelated to facility or equipment problems. Effective management can defuse this issue by planning in terms of safety-related maintenance items: tracking average time to correct items, initiating a reporting system on the status of maintenance items, etc. Other process-related targets include such indicators as: frequency of observations, percentage of people in a work group who are trained as Observers, frequency of safety meetings at which observation data is discussed, and number of observations done by managers.

ORGANIZATIONAL RESPONSIBILITIES FOR IMPLEMENTATION

From site manager to hourly employee, Implementation requires that people at all levels of the facility are clear about their responsibilities to the effort and to the following four functions of the behavior-based safety process itself:

- Planning
- Observation
- Facilitating Meetings
- Training

During the Assessment phase a clear assignment of roles and responsibilities is made to various levels and groups across the organization, and personnel are recruited to work as Planners, Observers, Meeting Facilitators, and Trainers during the Implementation effort.

Hourly Employees

Hourly workers often have a variety of views about safety, not all of them consistent. On the one hand, workers want to avoid injury; they want equipment and facilities to be safe; and they want co-workers who do not expose them to injury. On the other hand, workers also often view the enforcement of safety rules as a form of harassment—especially when enforcement is inconsistent. This can lead to an adversarial attitude in which hourly workers think of safety as management's responsibility and injury as management's fault. Insofar as 80%-95% of all injuries involve some unsafe behavior, it is clearly impossible for supervisors to directly prevent most injuries. This means that involvement of hourly employees is essential to the success of the behavior-based approach. This can be a delicate matter at first because hourly workers often hear the emphasis on changing unsafe behavior as a way of blaming them for injuries.

It is not a simple thing to establish an atmosphere in which workers are more open to the behavior-based approach. It helps to recruit hourly workers for participation in the planning, Implementation, and maintenance of the process; and an effective way of organizing this involvement is the use of an Implementation Steering Committee that has significant representation from hourly ranks. Issues in the use of a Steering Committee are discussed below.

Not all organizations are well prepared to involve hourly personnel in a significant way. Organizations managed in a traditional top-down manner find that neither workers nor managers are ready for a major safety effort requiring participation by hourly employees. Also, as important as hourly involvement is, it cannot be at the expense of management participation. A balance of the two is necessary, and management personnel, especially first-line supervisors, cannot feel that they have been ignored or bypassed. While the behavior-based safety effort is directed to hourly workers, it is important that first-line supervisors do not abandon the process and defeat it through passivity.

By assigning hourly employees significant roles and responsibilities during Implementation, management assures their involvement and participation in a way that is commensurate with their importance to success. By assuring the active involvement of hourly employees, the Implementation effort gains access to some of the best and most detailed information about safety-related behaviors at the facility. Hourly employees are often closest to the work, and their untapped knowledge represents a wasted resource. In the course of the Planning carried out by the Steering Committee, the behavior-based process analyzes and codifies this store of experience as the facility's inventory of critical safety-related behaviors.

Hourly employees with strong credibility among their peers are some of the best candidates for each of the four roles of the safety process: Planner, Observer, Meeting Facilitator, and Trainer. Many individuals become skilled in two or more of the roles. The behavior-based safety process not only depends on employee involvement, it is geared to employee involvement. This means that all the roles of the process are ones that hourly employees can excel at. Typical roles for hourly employees are:

- Steering Committee member, both at the departmental and the plant level
- Accident investigation, developing the facility's inventory of critical behaviors
- Observer
- Participation in safety meetings, engaged in identifying problems and doing problem-solving
- Trainer
- Facilitator

First-Line Supervisors

During Implementation, first-line supervisors receive training. In some facilities foremen and first-line supervisors support the activities of the Observers; in other facilities first-line supervisors do most of the observations themselves. In either case, since the first-line supervisors must be familiar with the observation process they usually receive Observer training, and they may also go through advanced Observer training.

Where first-line supervisors are responsible for conducting work group safety meetings, they also receive training as Meeting Facilitators—training that focuses on using observation data to identify problem areas and on communication and problem-solving skills. Effectiveness in these specialized skills presupposes training in the Foundation concepts of the behavior-based approach. And finally, once the continuous improvement mechanism is established at a facility, first-line supervisors are encouraged to do intensive work in special applications of the behavior-based approach. Managing the improved performance of employees with multiple accidents is an example of this kind of advanced work.

In facilities where first-line supervisors do not have responsibility for conducting work group safety meetings, they usually work with a team leader or team safety representative who is responsible for the meetings. In these cases the supervisor functions as a coach and as a resource to the team leader. Training in coaching for skills development can improve the communications skill and effectiveness of the supervisor.

In general the responsibility of the first-line supervisor to the Implementation effort is oversight in his or her area of supervision. The specific safety process roles and responsibilities of first-line supervisors vary according to the management style in their organization. In traditional organizations the first-line supervisor's function is very direct. In recent years, however, many organizations have redefined the supervisor's job description, away from *direction* and toward *providing resources* for the work group. In either case, first-line supervisors can function well in any of the four roles of the safety process.

In organizations where the first-line supervisor is primarily responsible for providing resources to the work group, it is the work group that is responsible for the Implementation of the behavioral process in its area. The first-line supervisor in this case is responsible for providing support for the process and for creating an environment which fosters employee involvement by doing such things as:

- Allocating sufficient time throughout the Implementation effort to see that necessary steps are taken.
- Encouraging designated Observers by giving feedback and consequences that are soon, certain, and positive in favor of consistent, timely Observations.
- Making sure that follow-up occurs for safety-related maintenance items.
- Ensuring that Observation data is used effectively in safety meetings.

- Assisting in the review of the facility's inventory of critical behaviors to keep it current.

In organizations where the first-line supervisor's function is very direct, supervisory behavioral process responsibilities may include:

- Doing Observations.
- Writing work orders and providing follow-up on safety-related maintenance items where there is no effective existing system for initiating and communicating about safety-related maintenance.
- Conducting safety meetings.
- Conducting sessions on problem identification and problem-solving.
- Accident investigation.
- Revising the facility's inventory of critical behaviors as needed.
- Overseeing proper Implementation of the process.

Middle Managers

Managers at the second level and above usually do not get involved in Implementation to the same degree as do first-line supervisors. At a minimum, however, these managers need training in the Foundation concepts of the continuous improvement process and in its practical applications. Though they are less involved than the first-line supervisors, middle managers are nonetheless encouraged to participate in the skill oriented training sessions on Observation, Feedback, etc. Their presence in such sessions has both practical and symbolic impact. In terms of practice, the more exposure middle managers have to the details of the behavior-based process, the more supportive they can be of Implementation. Symbolically, attendance by managers at Observer training sessions sends a powerful message about the importance of the safety effort.

Beyond the special considerations listed above, the middle manager's roles and responsibilities are much like those of the first-line supervisor but at a higher level. Middle managers make sure that their first-line supervisors are accomplishing their roles and responsibilities. Their duties as facilitators for their first-line supervisors may require that they draft, present, and pursue structural changes in the organization in order to provide an environment in which the first-line supervisors can perform their duties. It is worth emphasizing that for middle managers to give effective support to their first-line supervisors and foremen it is essential that the middle managers have a thorough understanding of the basic concepts, principles, and working mechanisms of the behavior-based process. This caliber of understanding requires specific training, not "general knowledge."

Site Manager

In facilities of up to approximately 200 employees, the site manager sponsors the Implementation effort. In large plants this sponsorship is the responsibility of the department or division manager. This means that the site manager not only knows the basic concepts of this upstream approach to safety performance, but clearly sees the scope of their application to the facility. With this overview, the site manager provides leadership and direction, encouraging the facility as a whole to endorse the behavior-based approach. The site manager is resolved to stay the course, communicating clearly that the resources required to make the process effective in the long run will be provided. The site manager must also be ready to pursue organizational development measures, especially in relation to his middle managers.

Middle managers may feel threatened by an approach that depends on such high levels of involvement by hourly employees. In addition, middle managers and first-line supervisors are often the hardest to get to buy into the benefits of the behavior-based approach to safety. These managers often mistrust any new initiative. They have become skeptical of the organization's ability to really improve. Based on history, their skepticism is often justified— after all, how did the word "program" get such a bad reputation? Middle managers and first-line supervisors are primarily responsible for balancing the pressure for production, quality, cost, and training. This makes them most vulnerable to a new initiative that is temporary—they put resources into it only to see it abandoned later. Middle managers are called on to make a judgment call as to whether the latest initiative is something that will succeed and last or whether it will fail and fall away. It is a difficult judgment call for them to make.

Sometimes an outside consultant can serve as the catalyst for productive discussion of these issues. An example is afforded by a chain of events at a chemical company in the Southeastern United States. At the company's facility, the authors conducted an Assessment very similar to the one presented in Part 2 of this book. The Assessment revealed that the middle managers and second level supervisors reporting to a certain department manager were basically skeptical about a successful Implementation effort. The facility had recently gone through an early retirement program and approximately two-thirds of the first-line supervisors had opted for retirement. Change at the facility was occurring at a rapid rate, and out of necessity middle managers were redefining their roles. They knew that the company wanted improvement in safety, but *Did they really have to do all this right now?* They were resisting the need for improvement.

Not wanting to make this situation more tense, when the consultant wrote up the company's Assessment Report he was reluctant to spell out this resistance. However, since it was a very important factor in the company's assessment, the consultant wrote a private letter to the department head, outlining the details of the situation in very straightforward language. To the consultant's surprise, when he next visited the facility in question, he found that he had been scheduled to address a meeting of the department head and those same middle managers whose resistance was of such concern to the consultant. The department head opened the meeting by reading the consultant's private communication to him aloud to the assembled managers and second level supervisors. The department head finished the reading of the letter by turning to the consultant and saying, "Now then, what did you mean by that?" The question put the consultant on the spot, but it was a very effective way to begin to address the real issues. What followed was a heart-to-heart talk about the Implementation effort that was scheduled to begin. The outcome of the meeting was that no one left the room that day with any uncertainty about his or her commitment to Implementation. This is not a recommendation of this particular method of facing up to facility issues. It is, however, an assertion that these issues present important opportunities for organizational development, providing that the sponsoring senior manager acts to address them squarely.

To ensure that middle managers and first-line supervisors are effective decision makers, the site manager provides the resources for them to receive thorough training in the basic concepts of the behavior-based approach. Once they understand the principles, the site manager makes it clear to middle managers and supervisors that decisions which compromise the safety effort are not acceptable.

THE STEERING COMMITTEE

Implementing and maintaining this process requires planning, good organization, and a time commitment. There are two primary ways to accomplish the task. A project team approach such as a Steering Committee can be employed or existing organizational structures such as the Safety Department may be used. Although each approach has advantages and disadvantages, most organizations opt for a Steering Committee since it allows for strong representation of a cross section of the facility, a factor which helps foster ownership of the Implementation effort at all levels.

There are some disadvantages to using a committee instead of an existing organizational structure; the committee may be less efficient. The people on a committee may be less accustomed to doing some of its required functions and so may do them less well or take longer to do them well. A committee usually takes longer to organize for smooth functioning and may require training in the skills needed to conduct an effective meeting. In addition, a committee presents logistical difficulties. Committee members may have to be released from their normal work assignments or else be paid for overtime work. If the organization has no recent history of using such project teams, a committee may pose real challenges. People often feel threatened by organizational structures which cut across normal lines of authority.

An alternative to both the safety department and to the project team approach is the facility safety committee. Most facilities have a safety committee of some kind, and it may be suitable for overseeing Implementation. An important consideration is whether the safety committee members can genuinely support the effort. Using an existing committee for the sake of convenience is no advantage if the members of that body do not support Implementation. Nothing is gained either if the safety committee is ineffective as a group due to political issues. Another consideration is the perceived status of the safety committee. If the safety committee members are not well regarded by their peers, it is detrimental to entrust the committee with direction of the Implementation effort.

The following description of responsibilities proceeds on the model of an Implementation Steering Committee. Facilities that do not use a Steering Committee, must nonetheless accomplish, by some other means, the same things that a Steering Committee accomplishes.

Planning, Communication, and Logistical Support

Planning, communication, and logistical support are the three major organizational tasks of Implementation. The Steering Committee typically handles all of these. In large plants the Steering Committee may be formed at the departmental level as well as plant wide.

Planning/Decision Making. It is most effective for a committee to do the planning and to make the important decisions of Implementation. Implementation means that change and acceptance of change is more likely when there is representative input from a cross section of the facility. When a Steering Committee is used for this purpose it is formed as early as possible so that it can contribute from the beginning of the Implementation effort.

Communication. One of the key challenges that the Steering Committee must effectively address is how to keep the entire facility and all levels of the organization properly informed and involved in every step of the Implementation effort. No group can be left out. The effective Steering Committee devises ways of reporting its own activities to the plant, of keeping the plant interested in ongoing developments, and of eliciting input from the plant when it is appropriate. The ease with which the facility accepts the behavioral safety process depends in large part on how well the Steering Committee communicates with the rest of the plant.

Logistical Support. Logistical support covers the nuts and bolts of Implementation. These include many aspects of training such as: preparing slides, giving talks and presentations, training Planners, Observers, Meeting Facilitators, and Trainers, making charts, etc. In some cases, members of the Implementation Steering Committee do virtually all of the support themselves. In other cases, the Steering Committee is primarily an advisory or decision making body, and the logistical support is done by specialists within the facility—the media department, the safety department, etc. When the Steering Committee relies on departmental resources for logistical support for Implementation, resource allocation also becomes a matter for decision making. Generally it is most effective to have the Steering Committee fill as many of the Implementation roles as possible. The extent to which this is possible depends on the individual and group capabilities of the Committee members—they may simply not have the necessary expertise. It also depends on the availability of other resources—no one else may be available to fill the Implementation roles, and it depends on how much time can be made available for the Committee members to work on Implementation—it may not be possible to draw people from their regular assignments for more than a very short period.

Key Decisions about the Committee

Reporting to the facility manager. The Steering Committee needs to fit into the organizational structure. Implementation requires phased resource allocation, which brings up multiple factors for consideration. It is rare that the Steering Committee can make the necessary resource decisions by itself. Usually its recommendations are presented for review and coordination. It is most effective for the Implementation Steering Committee to present its recommendations directly to the manager of the unit where Implementation is underway. Depending on the size of the facility, this is either the facility manager or the department manager. The reporting arrangement may include the manager's staff.

The reporting relationship has both practical and symbolic effects. The level at which the Committee reports communicates a message throughout the organization. There is beneficial symbolic impact in having the Steering Committee report directly to the facility manager. Because it is common for safety to be the responsibility of the safety department, whose head often does not report directly to the facility manager, a direct reporting relationship from Steering Committee to the facility manager can highlight the importance that is given to safety. In addition to the symbolic benefit, the practical advantage is that it is preferable to have Steering Committee recommendations immediately reviewed and decided on, and the level at which this can happen is usually that of the facility manager or department head (and staff).

On the other hand, where the Implementation Steering Committee does report directly to the facility manager, input from all levels of management is very important to success. For instance, if middle managers and first-line supervisors are inadvertently overlooked, they may see the Implementation effort as one that is driven principally by hourly employees. Middle managers and first-line supervisors need to have a voice.

The Steering Committee Facilitator. The person who facilitates the committee is clearly very important. The Facilitator must have the skill necessary to manage a large project and to work effectively with other people, especially in a project structure. The safety head is often thought of as a logical choice for the job. Whatever the merits of that individual may be, the safety head may not be the best choice. Having the safety head facilitate the Implementation Steering Committee may lead people to think that behavior-based safety management is just another program from the safety department. Someone else in the management position and who has the skills for the project can make a good Facilitator. However, a potential drawback to having a manager facilitate the Committee is that the hourly employees on the Committee might not participate as freely if a manager is the Facilitator. Having someone from the hourly ranks act as Facilitator of the Committee has powerful symbolic value. Acceptance by the workforce can be enhanced by having an hourly employee coordinate the Implementation effort. The possible drawback is that hourly workers have less experience in such roles. In this situation it is helpful to establish a coaching and advisory role between an appropriate manager and the hourly employee who is facilitating the Steering Committee. The Facilitator would report directly to the facility manager, and the lower level manager who is the coach and advisor is there to interact with the Facilitator about process issues.

The Steering Committee Facilitator has several important tasks. Concerning the Steering Committee/s, he or she attends liaison meetings and provides leadership and coordination. The Facilitator also must be able to assess the Implementation training needs of employees and to monitor the reliability of the Observers' data. In addition, the Steering Committee Facilitator does monthly reports, keeps Implementation moving, makes routine management presentations, and keeps up with consultant developments. In order to do these things well, the Facilitator needs to be versed in all four of the key functions of the behavioral safety process: Planning, Meeting Facilitation, Observation, and Training. For more on these functions and their related skills, see Chapter 10.

Selection of the Committee's Facilitator is done in time to allow him or her to have a role in selecting the other members. The pitfall for any Facilitator is that he or she will assume too much responsibility for the Implementation effort. This is not good for long-term success. These and related issues can be addressed at the outset by selecting a Facilitator who has good delegating skills, by arranging for some concentrated coaching for the Facilitator, and by giving the Committee a clear charter.

The Committee's Charter. To save the Implementation Steering Committee from floundering, it is important to provide structure from the outset. A charter is one way of doing this. Details of the charter will be modified as Implementation progresses, however it is best to anticipate the major issues and to make clear to all parties what the Committee will be responsible for in the way of results and timelines. An important issue to address explicitly in the Committee's charter is the relationship of the Committee to line management, especially first-line supervisors. It is best to develop a clearly defined procedure for having Steering Committee recommendations reviewed and endorsed by *all* levels of management.

Steering Committee size. There are so many variables of organizational culture and site that it is not possible to specify Steering Committee composition in detail. There are, however, some general rules. A single plant-wide Steering Committee is usually sufficient for facilities with fewer than 200 employees. At a facility with more than 200 employees, a single plant-wide Steering Committee may not be effective, and additional Steering Committees need to be established at the departmental level. Roles and responsibilities of each Steering Committee must be well defined.

Within the limits of efficient committee function, the number of people on the Committee depends on the size of the facility and on how much of the Implementation logistical support the members are expected to do themselves. At the high and low end, however, committees of this type rarely function well with more than 12 members or with fewer than 5. With fewer than 5 members, the work of the Committee can be too easily set back by unexpected absences, vacation time, and emergency work demands. With more than twelve members, the Committee grows unwieldly. The most effective Steering Committee involves from eight to twelve members.

In some very large facilities, separate Implementation Steering Committees are set up for different departments. In such cases it is important to have strong coordination between the Committees in order to maintain consistency of Implementation on key points. One coordinating approach at very large facilities is to have one Steering Committee at the facility level and subcommittees as needed by area. Each subcommittee is then facilitated or led by a member of the overall Steering Committee.

The single plant-wide Steering Committee generally has responsibility for the overall process and issues that are common to all departments. This oversight function includes:

- Outlining the fundamental elements of the behavioral process that each department must work toward
- Stating the objective methods for measuring departmental success
- Providing coordination and resources as needed
- General responsibility to make the process work

The departmental Steering Committee ensures that the process is working effectively at the departmental level. This includes:

- Conducting introductory presentations
- Developing the departmental inventory of critical behaviors
- Training Observers in the operational definitions and in Data Sheet use
- Skills development training for Observers and supervisors in verbal feedback, interviewing, managing resistance to change, and safety meeting participation
- Ensuring that Observation data is used effectively in safety meetings
- Incorporating accident investigation data into the inventory of critical behaviors and the Observer Data Sheet
- General responsibility to make the process work

In addition the department Steering Committee takes the pulse of Implementation effort to be sure that communications are clear and that hourly employees understand and support the basic process. During Implementation, communications issues almost always arise—incidents occur which are misunderstood or misinterpreted, rumors spread. The Steering Committee closest to these developments stays on top of them and acts to dispel rumors and to keep communications clear and open.

Composition and selection of the Steering Committee. The makeup of the Committee has a strong influence on where the ownership of the continuous improvement process will eventually reside in the organization. In order to work, the behavior-based process requires significant hourly involvement, and this means that hourly workers make up a high proportion of Committee members. Furthermore, as many important constituencies as possible need to be represented. For instance, there may be differences in values and practices—perceived or real—between departments or between shifts. To the extent that these differences are important at the facility, all of those organizational units need to be represented in some way on the Steering Committee. In a union facility, thought needs to be given to whether union officers are included on the Implementation Steering Committee. First-line supervisors and middle managers need to be represented. Given the membership constraints of 8 to 12 people, representative selection requires a balancing act. Broad representation is necessary, but even in some smaller facilities not all organizational units can be directly represented. When key organizational units are not represented on the Committee, it is necessary to devise ways of actively involving people from those units early in the Implementation effort.

Care is also given to how the Committee members are selected. The best people are the opinion leaders, people who are respected by their co-workers and whose endorsement will carry weight. Opinion leaders are not necessarily in formal leadership roles, however. Supervisors and managers usually have a sense of who the opinion leaders are. It is sometimes desirable to talk to candidates and sound them out before Committee membership selection is made. Committee participation tends to have more meaning for people when they feel that they have been selected through some degree of competitive process, and when they personally feel that safety is an important issue.

Tenure of Steering Committee Members. The heaviest time demand of Implementation is getting successfully through the Kickoff Meetings. The Steering Committee is very busy up through these meetings, at which the facility baseline is presented to the workforce, and the continuous improvement process is launched on an ongoing footing. It is misleading, however, for the Committee to focus solely on the Kickoff Meetings, because the quality of effort after those meetings also has a great deal to do with determining the long-term success of the effort. It is typical for the Committee members to experience a letdown after the observation process becomes more routine. To combat this slackening of the Committee's focus, it is usually a good idea to set Committee tenure at some period that extends past the Kickoff Meetings. Some facilities establish the Committee for an indefinitely long time, rotating the membership and the Facilitator on a predetermined schedule. The schedule assures people that they will not be trapped on the Committee, and it also moves people on who might otherwise resist leaving the Committee. The Facilitator needs to rotate so that safety does not become one person's pet project.

ROLES AND RESPONSIBILITIES OF THE BEHAVIORAL PROCESS

Implementation of the behavior-based safety process involves four functions: Planning, Observation, Facilitating Meetings, and Training. In addition, it can be very helpful for a facility's Implementation effort to have the benefit of an outside perspective. This subject is treated below as the Role of the Consultant.

During the Assessment phase, the assessment team is looking for likely candidates to fill the roles of Implementation. The most important organizational characteristic of the four safety process roles is that they are not exclusive of each other—one individual could serve as a Planner, an Observer, a Meeting Facilitator, and a Trainer. In practice, because of time and scheduling constraints, this usually does not happen. Nonetheless, the point is that there is nothing inherent in the roles themselves to prevent this—they are not hierarchically ranked, for instance. In fact, owing to the sequence of the Implementation effort, it is common for individuals on the Steering Committee to progress from Planning, to Observation, to Facilitating Safety Meetings. Once the Implementation phase ends, however, and the entire process is established as a self-regulating mechanism requiring all of these functions at the same time, most people specialize in one or perhaps two of these roles, rotating among them from time to time.

Throughout the Implementation effort Training is required. The ongoing behavior-based safety process requires a number of characteristic new skills and instruments. The Trainer's special skill is behavioral coaching for Skills Development, but he or she must also be fluent in the other skills of the process. Trainers are responsible for coaching the other roles in their respective skills.

The role of the consultant. During the 1980s the behavioral process was implemented in a large number of facilities in diverse industries. Assessing completed Implementation efforts shows that it is generally worthwhile for a facility to have an outside source to assist the Implementation effort. This source could be a consultant, a resource from corporate offices, or an employee from another company in a related industry, someone who has participated in a successful Implementation effort. The outside source is helpful in several ways.

1. Credibility. When the behavioral process is first being introduced, an outsider can have great credibility because the consultant speaks from experience in multiple settings of the same kind, and because he or she does not have any vested interests in any battles that may be underway at the facility. In some sense the consultant's perspective is neutral.

2. Fresh perspective. An outsider sees aspects of the culture of a facility that an insider takes for granted.

3. Avoids the pitfalls. An outsider experienced in Implementation knows and recognizes the pitfalls that decrease the probability of success.

Whichever of these sources of outside assistance is used, the facility is well advised to steer clear of anyone offering canned programs. With the behavior-based approach it is essential that a facility adapt, not adopt. Organizational development is the key criterion here. Genuine assistance from a consultant, or any other outside source, is assistance which the facility uses to *develop its own process.* This approach is superior to any off-the-shelf program for two reasons. Each Implementation is site-specific and must be tailored to the local culture in order to assure success. The organizational development approach to Implementation fosters ownership of the process—another essential ingredient of success. Of the various approaches which satisfy both the site-specific and ownership requirements, usually the Train-the-Trainer model works best. Trainers are usually members of the Steering Committee. The consultant works with the Steering Committee to implement the behavioral process, beginning with the introduction of the process to senior management and the assessment of the facility's Implementation strengths and weaknesses.

PART 3. IMPLEMENTATION

Chapter 10

The Implementation Effort

STEPS AND ROLES

Implementation of the behavior-based continuous improvement for safety by a facility requires *adaptation* of Foundation concepts to areas of performance identified in the Assessment. Since important differences exist even between facilities of the same company, in its details Implementation is as various as the facilities that undertake it. Nonetheless, there are some broad regularities and patterns for Implementation, and it is the aim of Chapters 10-16 to sketch these broad requirements of on-site adaptation in terms of the associated steps and roles involved.

Implementation involves a team of Implementers who carry out the four functions or roles of Trainer, Planner, Observer, and Meeting Facilitator. At its core, Implementation proceeds in the five steps that are presented below in Chapters 12-16. Briefly stated, the steps of Implementation are:

1. Develop the inventory of critical behaviors (Planner).
2. Present the inventory to a cross section of the workforce for review and endorsement (Meeting Facilitator).
3. Use the inventory to measure the baseline %Safe behavior of the facility (Observer).
4. Present the facility's %Safe baseline to the entire workforce of the target area—whether pilot area or site wide (Trainer, Meeting Facilitator).
5. Establish the continuous improvement mechanism on the basis of ongoing Observation and feedback (Planner, Meeting Facilitator).

Throughout the rest of this book, the names used to refer to the steps and roles of Implementation are not the important thing. The names are generic names, chosen to be as descriptive as possible of the functions that they refer to. The functions themselves are the important things. For instance, some facilities refer to behavioral Observers as Samplers. They use this name because the Implementers there discovered that their workers did not like the idea of being "observed." By contrast, sampling was for these same workers an unobjectionable practice that they were familiar with from quality improvement efforts. Since the Observer's function is to measure random samples of safety-related behavior, the name *Sampler* is also a clear and descriptive name of this function. The function is what counts, and the four key functions of Implementation are Planning, Facilitating Meetings, Observing, and Training. Planning is usually one of the key responsibilities of the Steering Committee. In addition, Steering Committee members may also serve as Meeting Facilitators, Observers or Trainers, but usually the Committee shares these responsibilities with other facility personnel. Before turning to a detailed presentation of each of these functions and their respective skills, a note on the data management system.

DATA MANAGEMENT SYSTEM

Once the behavior-based process is Implemented, there are two to three observations per work group per week. At the departmental level there could be hundreds of observations per month. Without a computerized system of managing the observation data, it is difficult, if not impossible, to analyze it effectively. In principle there is nothing to prevent manual collation and analysis of observation data. In practice, however, the work would be very cumbersome since in addition to straightforward plotting of %Safe figures, the observation data contains a great deal of valuable information whose availability and usefulness requires the fine-grained analysis that data base systems are designed to provide. Though there are many helpful and ingenious ways of analyzing observation data, the following are some of the more important analyses.

Frequency of Observations. The basic notion behind behavior-based safety management is that people change their behavior when they get feedback on their own operationally defined performance. Feedback needs to be frequent, especially in the early stages of the process. Observation also needs to be frequent and regular. At the work group level the frequency of observation can be read right off of the feedback chart. This data needs to be summarized, however, in order for the Steering Committee or for a department manager to monitor frequency of observation across work groups.

How often are specific inventory items Observed? In most facilities, a fairly small number of behavioral categories account for most of the injuries. For instance, either of the two behavioral categories of Body Position or Tool Use might be involved in one-third of a facility's injuries. These categories, however, might be only two out of six or more categories on the observation Data Sheet. The Observers might find it easier to notice items in the housekeeping categories of the Data Sheet, with the result that Housekeeping is observed very frequently while Body Position and Tool Use are observed only a small percentage of the time. This state of affairs is counterproductive. In order for the observation-feedback process to be effective in reducing injuries, attention needs to focus on behaviors that are more likely to serve as the common pathways to injury. It is necessary therefore to monitor the rate at which Data Sheet categories are actually being observed.

Safe behavior percentage by category. Unless overall safe behavior performance is very high, there is variability across observation categories—some of them being scored safe almost all of the time, others showing up with much lower percentages. At the least, the categories which consistently show low performance ratings may represent the areas in which workers are exposing themselves to injury. The categories with lower %Safe scores are the ones that the relevant work groups need to target for some guided problem-solving to discover why the unsafe behavior persists, and what kind of new consequences can be brought to bear to improve safety performance on those items.

Other analyses. In a shift-work environment it can be useful to analyze safe behavior percentages by shift or by time of day. In a rotating shift situation, crews may not differ in their injury rates, but rates may be different by time of day or by days on shift. For instance, there may be a higher rate of injury on the graveyard shift. If there are differing injury rates by shift, it is important to see:

A. whether the behaviors involved in the injuries are being observed during the critical shift; and if the answer to A is yes, and

B. whether the increased observation and feedback having any effect in improving shift safety performance.

It can also be helpful to compare Observers, within a work group, for instance. This important check of the consistency of observation across the group can be done by comparing Observer Data Sheets to see whether they score roughly the same categories with roughly the same frequency, and whether they arrive at roughly the same %Safe scores on the same categories.

Data Entry

The following information is entered into the data management system.

Observer identification. In order to compare Observers and to track the frequency of observations by individuals, unique identification is needed for each—a clock number, social security number, etc.

Organizational identification. It is useful to summarize the data at various organizational levels, from the work group on up to the entire facility. The work group needs a code that includes sufficient elements that can be summarized upwards.

Shift. In a rotating shift environment, the crew (or shift) needs to be identified.

Date. To summarize data over varying periods of time.

Time of Observation. If injuries are more likely to occur at certain times of the day, it is desirable to do observations more frequently at those times. Therefore the time of observation is important data.

Category and number of safe/unsafe behaviors. To track the safes and unsafes scored under the various categories of the facility inventory by the Observers.

Comments. It is a good idea to have a Comments Section on the Data Sheet. These comments are entered into the data base. If a category has been marked unsafe, this comment allows the Observer to specify exactly what was unsafe. For instance, if the category *Spills* was marked unsafe, the comment might note that there was a leak from a particular piece of equipment. If the category *Mobile equipment* is marked unsafe, the comment might be that a forklift was parked unattended with the forks not lowered to the ground. The Observers also use the Comments Section to record their thoughts about the antecedents and consequences that trigger and reinforce the unsafe acts that they observe.

Reports

Given the amount of information generated by the Observers and the number of variables available for analysis, personnel can easily be overwhelmed by reports. It is better to err on the side of simplicity, providing information that the work group can and will act on, than to present sophisticated analyses which will go ignored.

One of the keys to the success of behavior-based safety management is to provide a work group with relevant information about their performance. Problem areas can be identified, plans can be developed, and feedback about improvement will reinforce better performance. Safety meetings provide one of the better ways to get this information to the work group.

Whoever facilitates or leads the meeting (foreman, team leader, team safety representative, etc.), is in charge of assembling and presenting the information to the work group or team. This means that a report containing the basic data for the work group should be produced automatically and delivered to this person routinely. If the work group has a monthly safety meeting, the Meeting Facilitator should receive a monthly report summarizing observation data for the previous month. The Facilitator needs to receive this report several days to a week before each meeting. Information that does not get to the work group is information wasted. It also represents a break in the continuity of feedback, which slows the improvement process by making feedback less effective.

It is also desirable to produce standard reports regularly for various levels of management. In addition to standardized reports, the reporting system should have the capability of producing customized reports. A standard report might summarize data for the preceding month: a custom report might summarize data for the preceding three months. An elaborate reporting system would be able to summarize data by the preceding three months and display it with the comparable data from the preceding year, etc.

Data Flow

An orderly system for routing observational data needs to be established. Three main things need to happen with the information contained on the Observer Data Sheet. First, a copy of the observation sheet should be posted for a period of time adjacent to the chart on which overall %Safe behavior is plotted. The observation sheet itself gives considerably more information to the work group than does the overall number.

Second, information about maintenance or repair items has to get into the hands of the person who can do something about them. Usually this means that supervisors need to get a copy of the observation sheet so that they can fix the problem or take action to get it fixed (such as by writing a work order).

Third, the observation sheet has to get to the data entry clerk so that the information can be entered into the system. Personnel assigned to enter data vary widely depending on the circumstances at a particular facility. At some facilities these people have been secretaries, clerks, people on light duty assignments, and security guards.

At the time of Implementation, most facilities are already using data processing to monitor their rudimentary safety statistics. From the point of view of these existing data processing procedures, Implementation represents an increase both in the flow of data and in the sophistication of the generated reports and graphs. At some facilities the Implementers establish their own data processing staff. At other facility's the Implementers establish a liaison with the data processing department to handle the new behavior-based data. However these arrangements are made, the data base computer program that is used must have the flexibility and specificity to match the facility's inventory of critical behaviors, its Data Sheet, and its various needs for feedback reports, charts, and graphs.

THE FOUR FUNCTIONS OF THE BEHAVIOR-BASED SAFETY PROCESS

The four key functions of the behavior-based safety process are called on by the Implementation efforts in the following steps:

Step 1. *Planning* the facility inventory of critical behavior.
Step 2. *Facilitating Meetings* in which the inventory is reviewed.
Step 3. *Observing* the workforce to take a baseline measure of the facility's safety performance.
Step 4. *Facilitating Meetings* which present the baseline figures and the behavior-based safety process to the facility as a whole.
Step 5. *Planning* at the level of the work group, to establish the complete feedback loop of the safety mechanism.

At the last step, the behavioral safety process functions of Planning, Facilitating meetings, and Observing are on-line simultaneously, and the

The Four Functions of the Behavior-based Safety Process

PLANNING

Skills :

Analysis
- Cause Tree Analysis
- ABC Analysis
- Statistical Analysis of Injury Data
- Behavioral Analysis of Accident Reports
- Pattern Search
- Interpreting Observation Data
- Accident Investigation

Communication
- Managing Resistance to Change
- Investigative Interviewing

Products :
- Inventory of critical behaviors
- Observation Data Sheet
- Behavioral Action Plans

FACILITATING MEETINGS

Skills :

Analysis
- Interpreting Observation Data

Communication
- Conducting Group Problem-Solving
- Presenting Observation Data
- Giving Effective Verbal Feedback
- Managing Resistance to Change

Products :
- Safety meetings integrated into the behavior-based process

OBSERVATION

Skills :

Analysis
- Behavioral Observation

Communication
- Giving Effective Verbal Feedback
- Managing Resistance to Change

Products :
- Completed Data Sheets with % Safe ratings and special Comments
- Input on new behavioral inventory Items

TRAINING

Skills :

Analysis
- Trainee Performance Evaluation

Communication
- Behavior-based Coaching for Skills Development

Products :

Classes in
- Planning
- Observation
- Meeting Facilitation

Special Applications
- Self-Observation
- Managing employees with multiple incidents
- Behavioral Back-Injury Prevention

Figure 10-1.

safety mechanism becomes self-regulating and self-sustaining. The task that accompanies the Implementation effort throughout is Training. Steering Committee members need training in planning skills. Meeting Facilitators need training in how to present the Planner's inventory and Data Sheet for review, and Observers need training in how to use the reviewed Data Sheet to measure the facility's safety performance. See Figure 10-1 for an overview of these four functions.

Planning

The primary focus of the planning is to develop, and then refine, the facility's inventory of critical behaviors. This is both the first and 'last' step of the behavior-based continuous improvement process. In fact, since the improvement process is continuous it has no last step. Instead it forms a self-regulating loop or mechanism. Nonetheless, for introductory purposes it is useful to think of the process in a linear way. Furthermore, since the Implementation effort proceeds as a sequence of ordered steps it is helpful to see planning as the task that both opens and closes the loop of the continuous improvement process (see Chapters 12 and 16). Planning applies a number of analytic and communication skills to a facility's safety performance in order to produce the central instruments of the continuous improvement process:

* The inventory of critical safety-related behaviors
* The Data Sheet/s used by the Observers
* Behavioral Action Plans

Planning is one of the most demanding roles of Implementation because in addition to the analytic skills listed below, Planners are also ambassadors and troubleshooters of the new safety process. In these latter capacities they are called upon to exercise the communication skills of managing resistance to change and of interviewing. Planning relies primarily on several powerful analytic tools:

* Behavior-based Accident Investigation
* Behavior-based Accident Data Analysis
* Cause Tree Analysis
* Pattern Search
* ABC Analysis

Meeting Facilitator

Employee involvement is essential to the success of the continuous improvement process, and therefore meetings take on great significance. A properly focused and well conducted meeting is the most effective way to convey the scope of the process to a group of people and to engage them in its continuous improvement. The Implementation effort itself calls for meetings at two important junctures—inventory review, and baseline presentation, Steps 2 and 4 respectively. As the continuous improvement process matures, many facilities capitalize on the momentum of these meetings by developing them as an ongoing forum for performance feedback to the work group and process input from the work group. In high-functioning organizations, the work group safety meeting addresses many of the issues of planning as the group analyzes its own performance and identifies areas for improvement.

Observer

The primary functions of behavior-based observation are to measure workforce performance of the critical safety-related behaviors identified in the facility's inventory and to provide feedback on performance. Using the Data Sheet developed, reviewed and endorsed along with the facility's inventory of critical behaviors, Observers establish the facility's baseline safety performance. The length of time required for this third step of Implementation varies depending on the number of Observers to train and on the complexity of the facility's Data Sheet. After the initial Observer training, baseline observation periods typically last from two to five weeks. Establishing the facility's safety performance baseline does not require secrecy and so the Observers go about their duties openly. The fact that the workforce knows it is being observed does not adversely affect the baseline measure that the Observers produce. In fact, the high visibility of the Observers during baseline measurement has the important side effect of increasing employee curiosity about the new process, and therefore increases their involvement in Implementation.

By the time the baseline measurement is underway, the workforce in the target area knows something about the behavior-based process. Some of the workers have taken part in the Assessment interviews, surveys, and discussions. Others have attended the inventory and Data Sheet review and endorsement meeting/s held in preparation for training Observers in the use of these new instruments. The workforce may have been involved in accident and data analysis carried out during the planning which developed the inventory and Data Sheet. This background knowledge is all to the good because it makes the safety process Kickoff Meeting/s all the more credible and informative. It is at the Kickoff Meeting/s that the workers as a group learn the Foundation concepts of the behavior-based approach to safety and also find out how they measure up against the inventory of critical behaviors developed from the facility's distilled accident data and with the benefit of the experience of their most knowledgeable co-workers.

After the Kickoff Meeting/s, the Observers continue to make random samples of the facility's safety performance, but now they routinely engage in feedback with the people they observe. Facilities with ongoing safety meetings also depend on the Observers to participate in the work group discussions of observation data and feedback charts and graphs. In this respect, the facility's corps of Observers represents a valuable safety resource that is active in many ways other than the primary one of making safety observations. This is one of the reasons that it is very effective to train hourly employees to be Observers. They often make the best possible Observers because they are very familiar with the workforce and with the facility's equipment, processes, and products. They lend credibility to behavior-based observation by maintaining its integrity as a data gathering process rather than as a disciplinary procedure. Registering this important fact, some facilities train all of their hourly employees in these techniques and then rotate them through tours of duty as Observers. Other facilities train all hourly employees in observation whether or not they serve as Observers. In any case, in effective behavior-based safety efforts at least 50% of the active Observer corps is drawn from the ranks of hourly employees. This fact implies a great deal of training. The Trainer is active throughout Implementation, working with Planners, Meeting Facilitators, and Observers to develop the skills they need to produce their respective contributions to the success of the effort.

Trainer

Training is the universal function of Implementation. In Chapter 1 it was pointed out that at most facilities successful Implementation of the continuous improvement safety mechanism requires some degree of organizational development work. This is true even where SPC methods form the basis of quality improvement efforts. In other words, coaching in skills development is necessary in all roles and at all steps of Implementation. The training function is responsible for honing the skills of Planners, Meeting Facilitators, and Observers. There is even need at most facilities to train additional Trainers.

After Implementation is complete and the safety mechanism is in place in the facility, the training load decreases. However, as the workforce draws on replacements there are new Observers and Meeting Facilitators to train. Furthermore, once the facility's safety performance stabilizes in relation to the safety mechanism as a whole, various Special Applications often recommend themselves as fruitful areas for continued improvement. Some of the more typical Special Applications focus on employees who have had multiple incidents—oftentimes this relatively small number of workers has a significant impact on the performance of the facility as a whole. Some facilities employing a significant number of employees who work alone, for instance, often go on to develop an inventory and Data Sheet for Self-Observation for these and other solo workers. A facility may also want to use behavior-based safety principles to manage back-injury prevention or other special problem areas. For more on Special Applications, see Part 4.

Owing to the general importance of Training throughout the behavior-based continuous improvement process, training issues are presented by themselves in Chapter 11.

Chapter 11

Training for the Basic Skills of the Process

THE TRAINING MODEL

The perennial problem with industrial training is *transfer* from the training room to the work place. The training model used by behavior-based safety management addresses this problem on three fronts simultaneously.

1. Integrated ongoing system. Behavior-based training develops a process that is ongoing (identify→measure→feedback→re-identify) and integrates that system into existing safety systems. This approach provides a *structure* of activities, forms, procedures, and meetings that assures continuity of effort.

2. Skill-oriented training, in context. Behavior-based training coaches personnel in their roles and responsibilities *within* the developing process. Training does not happen in a vacuum. From the outset the training is structured for transfer from the training room to the work place.

3. Continuous improvement. Behavior-based safety management establishes a mechanism within the existing safety system for continuous upgrading of necessary skills—see *Coaching for Skills Development* in this chapter.

THE TRAINER'S ROLE

The Trainer is a key team member in the Implementation of the behavior-based continuous improvement process, see Figure 11-1. Usually the Steering Committee members act as Trainers during Implementation, having been trained in this function by an outside consultant. As the preceding chapter shows, the Planner, the Observer, and the Meeting Facilitator are primarily responsible for specific steps of Implementation. Trainers, however, provide coaching in the new skills of the process, and therefore their contribution is called for at all stages of Implementation. In addition to a thorough grasp of the behavior-based subject matter, the effective Trainer is both a coach and motivator.

Whether training Observers in the uses of the facility Data Sheet, or Supervisors in the techniques of interviewing, or of managing resistance to change, the Trainer's job is adult education. Since the trainees already have a great deal of experience about safety and human behavior, the Trainer's task is not so much to tell them something new as it is to get the trainees to think about familiar things in a new way. Furthermore, the training is for skills

TRAINING

Skills :

Analysis
- Trainee Performance
  Evaluation

Communication
- Behavior-based
  Coaching for Skills
  Development

Products :

Classes in
- Planning
- Observation
- Meeting Facilitation

Special Applications
- Self-Observation
- Managing employees
  with multiple incidents
- Behavioral Back-Injury
  Prevention

Figure 11-1. Training.

development. The Trainer does not simply convey information. Of course the new material is explained, but this is done as part of the process of helping the trainees to develop new skills. The Trainer provides the trainees with an opportunity to practice new skills in a structured environment. This skill-oriented training approach means that the Trainer is more like a coach than a typical classroom teacher. In addition to being good coaches, effective Trainers are good motivators. They introduce the trainees to a new skill, coach them while they practice it, and motivate them to use it on the job.

The following are some key places in the continuous improvement safety process where coaching may be needed.

Observers need training in behavioral observation, Data Sheet uses, calculating %Safe figures, charting %Safe figures, providing verbal feedback, managing resistance to change, interviewing and listening skills, and contributing to safety process meetings.

Safety Representatives typically need training in conducting group meetings, analyzing accidents, accident reports, Data Sheet charts and reports, summarizing and presenting behavioral data to safety meetings, listening skills, and conducting group problem-solving sessions.

First-line Supervisors. In addition to the tasks listed above for Observers and Safety Team Leaders, First-line Supervisors must be skilled at advanced steps of the process such as managing employees who have had more than one accident, teaching behavioral self-observation techniques, and measures such as behavioral back-injury prevention, see Part 4.

THREE BASIC SKILLS OF THE CONTINUOUS IMPROVEMENT PROCESS

The specific skills and responsibilities of the the Planner, Observer, and Meeting Facilitator are presented in following chapters as each one contributes to Implementation. There are certain basic skills of Implementation, however, skills which every key player uses. These basic techniques are the skills of managing resistance to change, and of investigative interviewing and listening. In addition there is the Trainer's own special skill of coaching for skills development. This is the skill that Trainers use in the conduct of their classes, workshops, and meetings. These three skills of Coaching, Managing Resistance, and Investigative Interviewing provide the foundation on which Planners, Observers, and Meeting Facilitators exercise their respective responsibilities of accident analysis, %Safe measurement, and performance interpretation.

COACHING TECHNIQUES FOR SKILL DEVELOPMENT

The behavior-based continuous improvement safety process requires new skills: observational skills, giving verbal feedback, presenting charts and data to a group, conducting group meetings, and interviewing and problem-solving. Helping workers develop these skills can be more difficult than it sounds. People learn at different rates and in different ways from each other. Some people find it difficult to admit that they need coaching to develop a new skill. The possibility of personality or learning-style differences between Trainer and trainee is always there. Effective Trainers follow a game plan that minimizes these difficulties by staying focused on skills development. In short, the Trainer exercises the skills of coaching.

Six Steps of the Coaching Process

Coaching is itself an important skill, one that deserves attention from the Steering Committee when they are considering roles and responsibilities of

Implementation. The elements of coaching are presented briefly as a six-step process, followed by a discussion of the salient points of each step.

1. Establish a coaching relationship. The Trainer establishes a coaching relationship with the trainees, a relationship that does not end until the trainees have achieved the level of skill they need for their designated roles and responsibilities in the continuous improvement safety process.

2. Identify target areas of performance. Together the Trainer and trainees select a target area to work on—observing behaviors versus seeing things, talking with a worker about observed improvement in his safety performance, helping a group to develop an effective safety action plan, etc.—and set goals for their performance of these skills.

3. Clarify the performance goals. The Trainer specifies very clearly for the trainee what good performance in the target area looks like. Trainers may model or demonstrate the performance level in question; they may show video tapes of successful performance of the skills. Above all the Trainer makes it very clear to the trainees what they are aiming at.

4. Provide for practice. The training involves at least two practice exercises for each identified skill. During the first exercise the trainees practice the new skill while the Trainer watches and makes notes of their performance. As soon as each trainee finishes this first practice, the Trainer gives *Success Feedback*—feedback addressed *only* to the areas where the trainee did well in his performance.

5. Subsequent practice. As the trainee prepares to give a subsequent practice performance of the new skill, the Trainer suggests a few things to work on and advises him or her how to make these performance changes. This is *Guidance Feedback*—feedback addressed to areas where improvement is needed. Once again the Trainer watches and makes notes, giving Success Feedback to the trainee as soon as the exercise is completed. This cycle of giving Success Feedback immediately after a performance and Guidance Feedback immediately before the subsequent performance continues until the trainee achieves the skill level that is needed.

6. Select a new target. The Trainer and the trainee repeat Step 2 above, focusing on a new target area for skill development.

Establishing a Coaching Relationship

Effective Trainers have credibility both in their subject matter and in their coaching manner. They are aware that their own particular learning style is just one among a variety of learning styles. The trainees feel motivated to improve their skills and they believe that the Trainer can help them. Since most people find it hard to admit that they are not as skilled at something as they would like to be, successful Trainers create an atmosphere where the trainees can seek improved results without first having to say that they are wrong. The Trainer makes an alliance with the trainee, expressing sympathy with the trainee's situation, *Well, you are really having to work hard to get the team behind the safety process—it must be pretty frustrating running those safety meetings.* The Trainers do not require the trainees to admit that they are not getting the necessary results. Instead they put themselves in the trainees' position in a sympathetic way. Then together they can simply focus on how to get better results out of the safety meetings. Throughout this first step of establishing a coaching relationship, a Trainer must be believable. The sympathy is genuine, and it is focused on skills development.

Identifying Target Areas of Trainee Performance

The Trainer has three primary sources of information about the trainees' performance: direct evaluation by the Trainer, the self-report of the trainee, and reports from co-workers. The good Trainer bears in mind that one observation is not enough—the trainee may be having an off-day. The Trainer is looking for areas of performance where the trainee is consistently below level. A trainee's self-report is often accurate and helpful—trainees are up against something unfamiliar and they know it. Reports from co-workers may be helpful, or they may provide very little help. The Trainer makes an independent judgment.

The Trainer and trainee then discuss what skill/s to work on first. The Trainer does not decide this matter alone, but gets input from the trainee, looking for the area/s of improvement that will make a significant difference and at the same time be relatively easy to accomplish. This first success builds confidence and supports the trainee's resolve to continue developing these skills. In cases where the most important improvement is also the most difficult skill to achieve, the Trainer may work with the trainee on something else first—it is important to have success at the beginning.

For the same reason, the Trainer and trainee work on only one or two skills at a time. Perhaps their targeted goal is to improve the way the trainee conducts safety meetings. They agree that when they finish the coaching the trainee will 1) speak more slowly and clearly, 2) address the back row and make good eye contact with the group, 3) organize his or her remarks and keep them brief and focused on the safety reports and charts, and in closing 4) direct the participants in productive problem-solving based on the observation data. This is too much for the trainee to keep in mind all at once, however. The Trainer works with the trainee on just one or two of these during any given practice exercise.

Clarifying the Goals—demonstrating good performance

The trainee needs a clear picture of what good performance is for each skill. It may be simply knowing the steps of Data Sheet routing, or how to interpret an observation data computer print-out. Discussion of such points may be sufficient. With other skills, such as conducting a safety meeting, giving verbal feedback, and other activities that involve interpersonal communication, however, watching someone else's performance is one of the most powerful ways to learn. The Trainer, or someone else, models the effective performance. The key here is that the model performance be good, not perfect. On the one hand, perfection is not very believable, and on the other, it can have the effect of discouraging the trainee. The Trainer makes sure that the trainees know what to look for in the demonstration. For example, coaching trainees in how to direct a group, the Trainer might suggest that the trainees pay attention to how the model talks to problem members, how he or she keeps the task before the group, etc. Another option is for the Trainer to model the good performance, taking over for the trainees at an actual safety meeting. This approach presents both advantages and disadvantages. On the plus side the trainees have an opportunity to see a demonstration in the very setting where they must do the same job. The drawback is that when the Trainer demonstrates it is harder for Trainer and trainee to discuss the performance later.

One solution is for both the Trainer and trainee to watch an expert model the good performance in the trainee's setting. The performance should be believable to the trainee. Video taped meetings offer the possibility of stopping the tape and discussing points as they come up, but it is hard for the camera to record all of the things happening in a meeting, and video tapes can be difficult and expensive to make. A group training session can be the answer to all of these points. The Trainer and trainees sit as members of a safety meeting conducted by someone who is good at it. The group demonstration assures uniformity of exposure to good standards. The trainees learn from each other's observations and from the Trainer's interactions with each of them. And of course when trainees do their first and then subsequent practice exercises, the others watch the performances and hear the Trainer giving feedback.

Success Feedback and Guidance Feedback

Separating Success Feedback from Guidance Feedback is important throughout the coaching process. The effective coach gives Success Feedback immediately after a trainee's practice performance and Guidance Feedback when the trainee is preparing for the subsequent performance. Since they are still learning to recognize good performance, trainees practicing new skills need to hear feedback immediately after they complete an exercise. Just before their performance exercise, the Trainer talks with them about the one or two things they are to work on. As the trainee does the practice exercise of the new skill, the Trainer takes accurate notes on the trainee's performance. When the trainee finishes, the Trainer gives Success Feedback on the particulars they have discussed — *Good performance. Much better eye contact this time, and I could hear you clearly in the back of the room.*

The Trainer saves suggestions for improvement until the next time the trainee does a practice exercise because the Trainer wants the trainee to take justifiable pleasure in the improvement that he or she has made, and because suggestions for continued improvement need to be made when they can actually guide the trainee's next performance. For this reason the Trainer gives Guidance Feedback to the trainees as they discuss what skills to work on in the course of the next exercise. Consulting the notes from the previous exercise, the Trainer says, *Good eye contact last time, and voice clearly audible — so during this practice keep those skills and concentrate on speaking more slowly and with longer pauses between your main points.*

Coaching—Summary

In its methods, behavioral coaching for skills development models the continuous improvement process. Performance areas are identified as opportunities for improvement. The Trainer watches and listens while the trainee performs the skills in question and arrives at an evaluation of the trainee's skill level — in doing this the Trainer assesses the trainee's skills baseline, so to speak. After that the Trainer gives the trainee opportunities to perform the identified behaviors more skillfully, providing positive feedback for success immediately after each practice exercise and suggestions for improvement before each subsequent exercise.

For the trainees to develop new skills they must change something that they already do. People naturally resist change, however. In their own work therefore the Trainers are often called upon to manage resistance to change. The trainees themselves also need to know this skill, because their new skills involve them in new roles and relations with fellow workers and colleagues. Implementing any new process means making changes, and, quite apart from the merits of the process in question, people are bound to resist the change. Managing this naturally occurring resistance to Implementation is therefore one of the underlying responsibilities of everyone who is engaged in implementing the continuous improvement process.

MANAGING RESISTANCE TO CHANGE

In addition to knowing how to coach trainees for skills development, Trainers are versed in managing resistance to change and in teaching others how to manage it. The continuous improvement safety process manages safety performance in advance of injuries and accidents. It does this by helping the workforce of a facility to increase the proportion of the time that they perform critical behaviors safely. To this end, Observers are trained, all employees become accustomed to giving and receiving safety feedback, and the organization finds systematic ways to interpret and use the data that their Observers provide. Each of these requires change both on the part of the individual workers and on the part of the organization as a whole. The incentive is injury reduction through continuous improvement, and improvement means change.

However much individuals and organizations differ in their responsiveness to change and in their resistance to it, some resistance to change is to be expected. In fact, resistance to change is an important part of human behavior. Far from being mere stubbornness or inflexibility, resistance to change is now known to be a positive force for continuity in human communities and associations of all kind. It is so much a part of our make-up that it follows patterns which can be described objectively and worked with for management purposes.

Generally the Steering Committee members, Planners, Meeting Facilitators, and Observers are trained in managing resistance to change. Resistance to change happens at the level of the individual and at the departmental and organizational levels. Observers learn to manage individual resistance to change, to prepare them for one-to-one and small group interaction and communication. Supervisors and Steering Committee members are trained in this as well and in addition are prepared to manage organizational resistance to change.

Managing Individual Resistance to Change

When people are asked to change, they offer a predictable response. Since resistance to change is a natural thing, dealing effectively with it can make the difference between a smooth Implementation of the continuous improvement process and a difficult and frustrating effort. The key word here is *natural.* Effective Trainers, Planners, Observers, and Meeting Facilitators realize that resistance to change is natural. They learn to see it in themselves, and thus they know not to take it personally. They learn to recognize it for what it is when they come across it, and therefore they are neither surprised nor distracted by it. They learn to manage it along with other factors of the Implementation of the behavior-based safety process. The facts about resistance to change are summarized in Figure 11-2.

Resistance to change is the opposition to change itself, not merely opposition to this or that particular change. This explains why people will resist even change for the better. The psychological function of resistance is to remove the demand for change, and it may be related to deep structures of

---

Facts About Resistance

A.   Some resistance to change is inevitable.

B.   The function of resistance is to eliminate the demand for change.

C.   Every person has his own automatic reaction to resistance.
1   It is important to know what yours is because it
will be a signal that resistance is occurring.
2   Usually these reactions are counterproductive because they often
involve giving in to resistance or resisting it. The first makes you
ineffective and the second is a fight.
3   Therefore, when resistance happens, it is important to
recognize it and not respond automatically.

D.   In and of itself, resistance does not mean a bad attitude.

Figure 11-2. Facts about resistance to change.

attention. In the same way that a constant or recurrent noise or background sound is filtered out, so perhaps resistance to change preserves us from being too responsive to demands for change and therefore incapable of continuing important projects and tasks. The most important point during the Implementation effort is to be aware that people are not bad employees or trouble-makers just because they resist change.

Resistance to change is natural because change itself is stressful to people. Postponing a long-anticipated vacation in order to paint the house is stressful. The vacation is something that is counted on. People count on their work routines too. Making a change may also imply that the previous situation was inadequate somehow. To agree with the need for change means that a person has to agree that the current practice has deficiencies, which is hard for some people to do. Change may also mean more work, even if it is just to learn new procedures. The new procedure may actually save time, but unless a person is convinced of the advantages of change the additional work may be seen as a burden only. The first day of a new promotion can be a stressful time, even though the promotion is a good thing. The newness of the situation is a change, and it represents the unknown. Even people who were quite skilled at their old jobs may worry that they will not be so competent in their new jobs. These are changes, and change is stressful (See Figure 11-3).

Why Change is Stressful

A.   Change entails additional work. The purpose of habit is to conserve
energy. The purpose of Resistance to Change is to protect habit.

B.   Change means facing the unknown. The unknown means insecurity
and anxiety. The function of Resistance to Change is to protect the
feeling of security.

C.   Change often involves giving up what seems right. People usually feel
that the way they are doing something is adequate. Resistance to
Change protects against capricious and unnecessary change.

Figure 11-3. Why change is stressful.

The Nine-Step Strategy of Managing Individual Resistance

The relation of change and stress has been studied, and the researchers
found that it makes no difference whether the changes are for the better—
the greater the rate of change in people's lives, the more susceptible they are
to illnesses and other effects of stress. Therefore effective planning incorporates and addresses these issues from the outset. The nine-step strategy for
managing resistance to change proceeds as follows:

1. Expect resistance.
2. Recognize it for what it is.
3. Be clear about what needs changing.
4. Communicate effectively.
5. Encourage discussion.
6. Ask for cooperation, not submission.
7. Have a realistic timeline for change.
8. Take suggestions and follow them up.
9. Be flexible and willing to negotiate the points that are negotiable.

1. Expect resistance. Resistance is normal. Like the basketball team whose
attention is not distracted from the ball even when individual players are
fouled, the Steering Committee is not distracted by resistance. They do not
take it personally. They follow the ball. If the Trainers are taken off guard, as

Common Resistances

Type of Resistance	Example	Mechanisms
A. Emotional	Anger, Guilt, Depression, Fear, Embarrassment	Produce an emotion to which you are "allergic" in order to make you back off.
B. Cognitive	Changing the subject. Irrelevant questions or objections. Failing to understand. Forgetting.	Sidetrack your thought, make you lose track of the issue, get you confused so that you will not be able to make an effective demand.
C. Social	*We're buddies, why get down on me?* ● *Why are you telling me how to do it — I've been doing it this way since you were in diapers.* ● *Your boss doesn't even do it the way you want me to do it.*	Alter the nature of the relationship between themselves and you so that it is inappropriate for you to ask them to change.
D. Behavioral	Fighting. Saying OK, but not really complying.	Threaten you so you will back down. Give the appearance of cooperation while resisting.
E. Psycho– Physiological	Being too tired. Getting sick.	Justify not changing on the basis of health needs.

Figure 11-4. Common forms of resistance to change.

soon as they realize they have been taken by surprise by resistance, they get back on track with whatever comes next in their duties and responsibilities. The same is true for the Planner, Observer, or Meeting Facilitator.

2. Recognize resistance. Implementers who recognize resistance are in a position to prevent their own automatic response from interfering with their capacity to foster cooperation. They use their own automatic responses to resistance to alert them to its presence. In this way they avoid fruitless power struggles. Psychologists identify various kinds of resistance: cognitive, emotional, social, behavioral, and psycho-physiological, see Figure 11-4. Implementers who effectively manage resistance to change learn to recognize these five kinds of resistance in themselves and in others. *Cognitive resistance* uses the thought processes to eliminate the demand for change and is probably the most common kind of resistance in the work place. Examples of cognitive resistance are such patterns as changing the subject of conversation, bringing in irrelevant considerations or objections, forgetfulness, and saying one thing but doing another. *Emotional resistance* works by making an emotional situation that the person asking for change is repelled by or "allergic to." Anger, guilt, and fear are often used in the service of resistance to change. *Social resistance* uses social roles to eliminate the need for change. Common strategies of social resistance involve turning the boss into a friend or the Observer into an inexperienced greenhorn, a nuisance, or a spy.

3. Be clear about what needs changing. Implementers who recognize resistance to change act to minimize it by removing the barriers to change. They have thought through the question of what needs to change and in which order. They have a clear picture of what the changes mean for the workers both in the way of advantages and disadvantages and they are prepared to present the advantages and to alleviate the disadvantages by providing resources where they are needed.

4. Communicate effectively. This means motivating the workforce in terms of the benefits of the change by showing them how it extends their existing goals and values—injury prevention, objective assessment of observable factors instead of personality differences, accountability, problem-solving by the work group for the work group, etc. The Implementers' primary point here is that the change to continuous improvement in safety is not itself negotiable—that is one change that is necessary. Workers must pay attention to the practices identified in the facility's inventory of critical behaviors. Having been clear about their expectations on this point, the Implementers go on to remove the barriers to change by explaining the process, its application, the history of its development at the facility, and in general giving as much detail as the workers find helpful. They are prepared to outline a step by step plan for the change.

5. Encourage constructive discussion. Wherever it is appropriate the Implementer encourages discussion to enhance understanding—this is not discussion that challenges the need to change the safety culture.

6. Ask for cooperation, not submission. The behavior-based approach to safety lends itself to asking for cooperation and not submission. The focus of this approach is on objectively observable behaviors. Therefore there is no need to tell anybody what to think or feel (attitude) about the safety process or even about the critical behaviors on the facility's inventory. The Observers get cooperation by being cooperative. They solve problems and find resources (part of removing barriers to change, also), and they put themselves in the position of the other employees. Employee involvement is essential to the continuous improvement mechanism of work group initiated problem identification and problem-solving. This kind of involvement cannot be imposed.

Submission is relatively easy to achieve in the short term, but over the long term it creates problems. It engenders resentment and therefore creates more subtle resistance. It also means a more negative atmosphere, one that takes resources and time for policing activity. Finally, submission makes people dull and slow-witted because it requires that they shut off their minds and obey whether or not they understand. Cooperation, on the other hand, makes people happier because they cannot cooperate without thinking and understanding, and they learn from each other as they work. It also decreases the natural tendency to resist change because in the cooperation itself there is mutual support for people during the change.

7. Have a realistic timeline for change. The Implementers indicate that they know change takes time and that they do not expect it to happen overnight.

8. Take suggestions and follow them up. Workers will have suggestions for how to make the change easier, quicker, etc. These suggestions need to be noted and followed up.

9. Be flexible—negotiate the negotiable. This follows from cooperative problem-solving. For the sake of completeness and good planning at the outset, the Steering Committee sketches a number of details that are just that, details, Not every facet or timetable of the changes is necessary or even critical. Effective Implementers are flexible about such things; they are willing to negotiate on them for the sake of the central, non-negotiable part of the process—accident prevention and continuous improvement.

Four Factors of Organizational Resistance to Change

Just as individuals may resist change, so may organizations. The organization in question may be an entire company, or it may be sub-units of a company such as a division, a plant, a department, shift, or work group. Strictly speaking, since organizations do not exist apart from the individuals who compose them, organizational resistance to change is still individual resistance. There are, however, some important factors of resistance which emerge at the organizational level. Four of these are cohesion, interdependence, territoriality, and management systems.

Cohesion refers to the way that individuals in a group reinforce each other. The more the members of a group agree on resistance to change, the more difficult it is to change them.

Interdependence refers to the way that a properly functioning organization has parts which act in a coordinated way. Where change is concerned, if one part of an organization resists change, its resistance is magnified because it interferes with the ability of other parts of the organization to accomplish change even though they are not resisting the changes themselves.

Territoriality refers to the fact that the change may threaten a unit's relative status within the organization as a whole.

Management systems have an important bearing on the ease with which change can be accomplished in an organization. Implementation of change is affected by such questions as: Who sets goals for whom? How much input into the change process do various units have? How persistent about implementation of change has management been in the past? What is the turnover rate of the facility's senior management?

Three Forms of Organizational Resistance to Change

The most common forms of organizational resistance are cognitive, minimal compliance, and out-waiting the change.

Cognitive resistance at the organizational level takes the form of endless requests for more information or interminable arguments about details before being willing to cooperate fully.

Minimal compliance is most common where a unit works in isolation from others, or when there is strong cohesion in the unit. This form of resistance often takes the form of satisfying the letter of changed procedures but resisting at every point the spirit of the changes.

Out-waiting the change is common where there has been a history of programs that come and go with little or no management follow-up, either because management itself turns over rapidly or because it lacks the will to accomplish long-term change. Out-waiting the change is probably the most common form of organizational resistance to change.

Managing Organizational Resistance to Change

The strategy for dealing with organizational resistance to change is similar to the strategy used with individual resistance. The first steps are simply to expect resistance and to recognize it when it happens. The Implementer makes cohesion work in favor of change by winning the support of the spokesman of the cohesive group (sometimes referred to as the Champion or the Hero of the group). If the group's spokesperson cannot be won over to the change, the Implementer may isolate him or her from the group so that it can consider supporting the new safety process without opposition from its own leader. Or the Implementer may work with the entire group at one time, engaging group cohesiveness in commitment to the training sessions.

Interdependence is managed by identifying and winning the help of key players in each group. This group of key players is made up of all of the people who need to cooperate to make the change a success. Sometimes this requires training both of the interdependent groups. The strategy is that both interdependent groups must be included in the change.

Territoriality is addressed by introducing the change so that key players have input into planning for the change and in shaping the change process itself. The concept of *shaping* is an important one for the managing resistance to change. In his book *Theory Z*, William G. Ouchi demonstrates that the more group involvement there is in shaping a change, the less resistance there is to the change.

Management systems that are capable of implementing long-term changes have key players who simply begin to act as though the change is here to stay. They do this because their implementation roles are clearly defined and they are committed to performing their roles visibly and consistently. Necessary support and resources are identified and allocated to these key players. And accountability for Implementation performance is built into the Action Plan.

INVESTIGATIVE INTERVIEWING TECHNIQUES

Effective investigative interviewing rounds out the basic skills of Implementation, complementing the skills of coaching and of managing resistance to change. Coaching for skills development provides trainees with structured practice in new procedures and measures. Techniques for managing resistance to change keep the new procedures and measures on track. Investigative interviewing provides the Implementers with important data and feedback from the workforce throughout the Implementation effort. Gathering and responding to such information comprises an important part of the responsibilities of Trainers, Planners, and Meeting Facilitators.

As the name implies, investigative interviewing is in-depth interviewing. Trainers use it to discover in which areas their trainees really need to develop their skills. Planners use investigative interviewing in one-to-one situations during either accident investigations or during accident data review. During the course of safety meetings and other small group gatherings, Meeting Facilitators use these interviewing techniques as part of brain storming and problem-solving exercises. Finally, investigative interviewing is central to managing employees who have had multiple incidents. For a detailed presentation of this work, see Chapter 19. Clearly these three situations differ from each other in important respects. Nonetheless, there are some essential similarities among them, and these similarities form the core of the skill of investigative interviewing.

Starting the Interview

Set aside dedicated time. Unlike casual or informal conversation, investigative interviewing requires dedicated time. The interviewer takes the responsibility of setting up a situation with minimal interruption and distraction. The purpose of this kind of interview is to discover important safety-related facts and insights, facts that help the Trainer to compose a curriculum, or the Planner to formulate an objective definition for the facility inventory of critical behaviors, or the Meeting Facilitator to guide the work group problem-solving effort. This kind of interviewing cannot be done while either party is being called to the phone, etc.

Non-judgmental, non-disciplinary. Behavior-based investigative interviewing is a win-win situation, not a zero-sum exercise in assigning blame. The interviewer assures the interviewee/s that the sole purpose of the interview is fact-finding. Once again, this point goes back to one of the most fundamental concepts of behavior-based management, namely that many conditions and actions precede any particular incident or state of affairs. The Trainer has no interest in blaming the trainees for needing training—the Trainer just needs to know how best to coach the trainees in the necessary skills. The Planners drafting a new category or item for the facility inventory of critical behaviors have no interest in blaming an injured worker for the injury—they just want to make sure that the facility's Data Sheet is accurate enough on this point that the Observers can measure the facility's overall behavior in relation to it. The Meeting Facilitator has no interest in blaming the work group for their observed low safety performance on some item/s of the facility Data Sheet—he or she just wants to discover *with them* how to improve their performance.

Manage resistance. This kind of interview discovers new things. It is also a new experience for many people, and therefore represents a change. Interviewees—trainees, injured workers, and work groups—may resist the interview process in any of the ways presented in the preceding section of this chapter on managing resistance to change. Keeping resistance management skills in mind, the interviewer stays on track and exercises the interviewing techniques of dialogue and active listening.

Effective Dialogue

The interviewer engages the interviewee/s in effective dialogue by:

- Asking open-ended questions
- Listening for promising leads
- Exploring contradictions in the interviewee's answers
- Eliciting all pertinent details
- Noticing changes of subject
- Checking for accurate understanding

Ask open-ended questions. The interviewer starts with open-ended questions and gradually focuses on details. An open-ended question is one that has no simple answer but requires some explanation. *Will you tell me about your accident?* is an open-ended question. *You have a bad back, don't you?* is not an open-ended question. Simple questions tend to be answered in an automatic or unthinking way; whereas in order to answer an open-ended question the interviewee has to think about the question and about his answer. This process allows the interviewer to learn more about the causal connections involved in subject matter of his question.

Listen for promising leads. Alert interviewers listen carefully, actively sifting what they hear for clues, contradictions, things that do not add up, etc. When such fruitful leads arise the interviewer follows them, restraining the interviewee from wandering too far away from the subject of the question.

Explore contradictions. The interviewer starts with open-ended questions and follows up the answers, narrowing down to specific points, gradually pursuing all of the clues identified in the interviewee's answer to the open-ended questions. The goal is to locate and analyze contradictions. The interviewer asks himself whether the answer really makes sense. If the answer contains contradictions then the interviewer cannot understand it. For example, a worker might say that she did not wear eye protection because it is *uncomfortable* and because *it did not occur to her.* How can both of these answers be true at the same time? The interviewer pursues this matter. The intent in this case would be to discover which consequences are actually driving the safety-related behaviors of the interviewee. Before this can be accomplished, however, the interviewer needs to discover just what those behaviors actually are. Perceived discomfort of eye protection calls for very different new antecedents and consequences in favor of safe behavior than does absent-mindedness or ignorance about the need for eye protection.

Elicit details. The interviewer does not stop asking for details until he or she has a clear picture of what happened, asking for all pertinent details. Effective interviewers also pay attention to their own interpretive habits and notice whether they are having to fill in any details in order to make the picture make sense.

Notice changes of subject. The interviewer pays close attention when the interviewee answers an unasked question, or does not answer the question that was asked, or changes the subject in mid-answer. Often this behavior signals an area of confusion for the interviewee. The interviewer brings the dialogue back to these areas and goes over them carefully with the interviewee.

Check for understanding. The interviewer restates the main points that emerged from the interview, checking with the interviewee to be sure that the interview notes are accurate.

Effective Listening

Listening is the other side of effective investigative interviewing. Many of the characteristics of good listening are the flip-side of good dialogue skills. Just as important as the dialogue skills listed above, good listening skills allow the interviewer to receive valuable information and to remove barriers that make people work at cross purposes. Most importantly, effective interviewers do not take listening for granted and they do not underestimate the carefulness required by alert and responsive listening. Inexperienced managers sometimes feel that listening is incompatible with supervising. On the contrary, however, good listening is a prerequisite to good supervising. This is because following what other people say does not mean agreeing with them in every case. It means understanding what is said. Good interviewers come to an understanding of what the interviewees say, whether they agree with them or not. Effective listening is:

- Non-judgmental
- Addressed to common ground
- Slow-witted
- Persistent
- Friendly
- Responsive
- In charge

Non-judgmental. In either case, whether the interviewers agree with their interviewees or not, they do not condemn or judge them. Investigative, behavior-based safety interviewing is not even remotely similar to a criminal investigation, for instance. The difference is that both parties share common ground.

Address common ground. The safety interview is a fundamentally cooperative venture, a fact which the effective interviewer is always mindful of. The Trainer is trying to find out in which areas the trainee could best use some coaching. Or the Planner is working with the employee to analyze an accident in order to make a plan to prevent just such accidents in the future. Or the Meeting Facilitator is encouraging the work group to brain storm the causes of its low safety performance on some item of the facility Data Sheet in order to do some problem-solving around the issue of how to improve performance.

Slow-witted. Especially good listeners say that the secret of their success is that they are not very bright. What they mean is that to be good listeners, interviewers must be slow-witted enough not to assume they understand what most people think is obvious. They question the obvious. This is a valuable strategy because oftentimes things do not make as much sense as people think they do. Investigative listening is comparable to searching for a lost treasure. The interviewees are the source; they know where the treasure is, but they are not aware that they know. They know, for instance, the range of consequences which determine whether they wear protective equipment, but they have never thought the matter through explicitly and clearly. Or they know exactly how to use a piece of production equipment safely, right down to the smallest tricks of the trade, but they have never thought to say this cluster of safe behaviors as an operational definition that an Observer could see or a Trainer could use. The listening skill of the interviewer is to be sufficiently slow-witted to elicit this kind of answer from the interviewees.

Persistent. Patience and persistence characterize good listening. Effective listeners go slowly, and when things do not add up they do not act as if they do. A hurried manner invites off-the-cuff answers. The patient interviewer makes it clear that he or she wants the detailed facts. A fast pace may hear what the interviewees think, but it does not encourage them to think the matter over in a new way or to see it from another angle. This fresh angle of approach, however, is precisely what both parties need in order to arrive at a new understanding.

Friendly. The effective listener takes time to be friendly. This helps to establish an alliance with the interviewee/s. Without this friendliness, the interviewer's good listening behavior may seem offensive. This slow-wittedness might be mistaken for skepticism, and the persistence for criticism.

SUMMARY

Coaching for Skills Development, Managing Resistance to Change, and Investigative Interviewing—these skills are sufficiently generic that they are called for throughout the Implementation effort, and beyond. They also find application in areas other than safety. One of the beneficial side effects of the behavior-based Implementation effort is the general enhancement of communications and analysis skills experienced by the participants. During Implementation one of the most important applications of these skills comes with the development of the facility's inventory of critical safety-related behaviors. Chapter 12 addresses this subject.

Chapter 12

Developing the Inventory of Critical Behaviors

Acting as Planners, the Steering Committee develops their facility's inventory of critical safety-related behaviors. The inventory is produced from many different source materials, and a glance at Figure 12-1 shows some of the skills and products covered in this chapter. The Planner uses several analytic skills to produce the inventory with its operational definitions of critical safety-related behaviors, the Observer Data Sheet, and behavioral action plans.

In 80%-95% of safety incidents, human behavior is the final common pathway leading to injury and accident. It follows that each facility—characterized by particular production processes, products, and workforce—has a characteristic cluster of these final common pathways which are responsible for a significant percentage of its safety incidents. The task is to identify this cluster of safety-related behaviors, write operational definitions of them to form the inventory, and then prepare a Data Sheet for Observers to use.

The preliminary stage of this work begins during the Assessment phase, see Chapter 6 on Behavioral Analysis of Accident Reports. The Assessment team reviews accident reports in terms of generic behavioral categories to find out which of them is responsible for significant numbers of incidents. These generic behavioral categories include such things as *Body placement and position in respect to task, Equipment use, Personal protective equipment,* etc. The purpose of that preliminary work is to demonstrate to management the cogency of the behavior-based approach to safety, to build support for Implementation, and to identify fruitful areas for more thorough analysis during the development of the behavioral inventory.

Since the approach measures carefully specified behaviors at a particular site, the inventory cannot be imposed by one department or facility on another. There is no substitute for developing an accurate and functional inventory of critical behaviors, although some companies have tried to find a substitute or shortcut. The most common shortcuts are to limit facility representation in Planning meetings, or to adopt a critical behaviors inventory from another site, or both. The facility that does this expects to save time, but it actually incurs a net loss because it loses the benefits of having an inventory of its real situation. The cornerstone of the behavior-based safety process is the inventory that identifies the site-specific behaviors that are critical to safety performance at the facility. Build poorly, and what follows is

```
┌─────────────────────────────────────┐
│                                      │
│  PLANNING                            │
│                                      │
│  Skills :                            │
│                                      │
│  Analysis                            │
│  ● Cause Tree Analysis               │
│  ● ABC Analysis                      │
│  ● Statistical Analysis of           │
│      Injury Data                     │
│  ● Behavioral Analysis of            │
│      Accident Reports                │
│  ● Pattern Search                    │
│  ● Interpreting                      │
│      Observation Data                │
│  ● Accident Investigation            │
│                                      │
│  Communication                       │
│  ● Managing Resistance to            │
│      Change                          │
│  ● Investigative Interviewing        │
│                                      │
│                                      │
│  Products :                          │
│                                      │
│  ● Inventory of critical             │
│      behaviors                       │
│  ● Observation Data Sheet            │
│  ● Behavioral Action Plans           │
│                                      │
└─────────────────────────────────────┘
```

Figure 12-1. Planning.

poor. Even more importantly, simply adopting someone else's inventory sacrifices employee involvement and participation. Employees who have input into the development of their own inventory are sold on it. This is because they get a chance to *see* the critical behaviors that are common to their various injuries, and in seeing this they recognize the inherent validity of the behavior-based approach to accident prevention.

SKILLS AND PRODUCTS OF PLANNING

The development of a behavioral inventory draws on various skills, both of analysis and of communication, to arrive at clear, objective definitions of a facility's most important safety-related behaviors. Such definitions are known as operational definitions because they are framed completely in terms of how something is done. This set of definitions provides the facility Observers with a resource that tells them exactly what behaviors to look for when they are taking samples of work place safety performance. The definitions, and their related categories, must be clear enough for the Observer to be able to recognize safe and unsafe behavior without interpreting a situation or guessing at someone's attitudes, alertness, and other internal states.

Behavioral Analysis of Accident Reports—Identifying Critical Behaviors

There are four ways to identify the critical behaviors of a facility:

- Analyze incident reports
- Interview the workers
- Observe the workers while they are working
- Review work rules, job safety analyses, procedures manuals, etc.

Of these four approaches, behavioral analysis of accident reports is the most important and should be completed first. This step provides the basic foundation of behaviors to target. Various mixes of the other three approaches may then provide useful supplemental information.

The best strategy for identifying a facility's critical behaviors is a thorough behavioral analysis of its accident reports for the period of the past three to five years. The Planners review these reports to discover which safety-related behaviors are associated most often with the facility's injuries. They then develop an operational definition for each of these critical behaviors, and the behaviors are listed as items on the Data Sheet. Typically the review process is quite painstaking at first because the patterns of behavior that the Committee is looking for are not documented very well by traditional accident reports. This means that the analysis must be carried out by employees who are familiar with the work. With the right techniques, the underlying patterns begin to emerge—the patterns of behavior that are actually driving the safety performance of the facility. The inventory of critical behaviors is an invaluable asset for preventing accidents and injuries. It also serves as a training guide for new employees.

Table 12-1 shows the Pareto Analysis of three years of behavioral data from a facility. The Planners have discovered that the first three critical behaviors alone have been involved in 47% of their facility's incidents in the three-year period they analyzed. The first five critical behaviors were involved in 70% of the facility's incidents. Notice that these critical behaviors are not the same as the typical "causes" coded on most accident report forms. Figure

Table 12-1. Pareto analysis of accident reports.

CRITICAL BEHAVIORS	% OF INCIDENTS
Eye and face protection	19
Body placement	14
Hand protection	14
Tool use	12
Line of fire	11
Foot protection	10
Pre-job inspection	9
Breaking-into procedure	5
Tool condition	5
Visibility	4
Confined space	3

Behavioral Inventory Categories & Items

			%	%
1.0	Procedure			
	1.1	Confined space	3	17
	1.2	Breaking-into process	5	
	1.3	Pre-job inspection	9	
2.0	Personal Protective Equipment			
	2.1	Eye and face	19	43
	2.2	Hand	14	
	2.3	Foot	10	
3.0	Tools			
	3.1	Condition	5	17
	3.2	Use	12	
4.0	Body use			
	4.1	Placement	14	29
	4.2	Line of fire	11	
	4.3	Visibility	4	

Figure 12-2.

12-2 shows the next step after Pareto analysis. The items are put into categories, and the category percentages are calculated.

After Implementation has established the continuous improvement process in the facility, these same techniques of identifying critical behaviors are used during accident investigations. Most facilities that the authors have worked with have thoroughgoing procedures for accident investigation. The added cogency and penetration of the behavior-based approach to accident investigation is that its focus throughout is to discover the critical behaviors that were involved in the accident, along with important management system issues and information about facility equipment and conditions. The categories of the facility inventory and the observation Data Sheet guide the investigation. Where the investigation discovers items that are not yet in the inventory, this blind spot is remedied. In other words, accident investigation becomes an integral part of the ongoing Data Sheet development loop at the facility.

Identifying the Critical Behaviors in an Incident

The three examples below were taken from incident reports. Each is followed by a brief analysis into critical behaviors and related inventory categories. During inventory development, the Steering Committee members engaged in Planning read through their collected incident reports, identifying the critical behaviors associated with the incidents.

Incident 1. As a mechanic was breaking into a steam line, his helper was burned on the face by hot water escaping from the flange that the mechanic was opening.

Critical Behaviors	Inventory Categories
Avoid line of fire	Body use
Head/face protection	PPE
Breaking-into process	Procedure

Incident 2. While he was using an open-end wrench to loosen bolts on a flange, the worker's hand slipped, and he cut his forearm on a sharp corner of a machine guard. The machine guard—which was not directly visible from the work location—should not have been sharp.

Critical Behaviors	Inventory Categories
Do not work where you cannot see	Body use
Use only proper tools	Tool use
Inspect work area for hazards	Procedure

Incident 3. While he was attempting to close a 6-in. valve, a chemical operator stepped three feet above ground and leaned over two 1½-in. lines. The "cheater" slipped, causing the operator to fall forward and resulting in a pinch laceration to his hand.

Critical Behaviors	Inventory Categories
Body placement both unstable and extended	Body use
Using wrong tool	Tool use
Wear gloves	Procedure

Cause Tree Analysis of Accidents

Most generally speaking, accident analysis indicates what there is to learn from past accidents about their causes. In addition, Cause Tree Analysis of accidents serves several important functions. It represents an impartial, non-blaming method of analysis. Cause Tree Analysis is exhaustive, clearly laying out all of the causes as they contributed to the accident—identifying human causes along with the equipment, process, and procedural causes as well. This approach arises from the basic premise that any accident is the last step of a chain of causes and more typically of several chains of causes. Cause Tree Analysis untangles these chains of causes and shows each in its separate role in the accident in question. This clarity of analysis is necessary. Unguided by such a method, many people lump causes together, either to dismiss them or to assign blame. The tendency is either to focus on the workers, blaming them for being inattentive, or on the company, blaming it for faulty conditions. This reactive tendency does not help to reduce injuries.

Figure 12-3 shows a Cause Tree Analysis form and Figure 12-4 a completed analysis for an accident in which a sample chemical splashed in the eye of a worker. Accidents usually involve two or more things coming together in the wrong place at the wrong time: in this case, the worker's eye and the chemical sample. On the top line the injury is written in, *Sample splashed in eye.* On the second line there are spaces for recording the *immediate* causes of the accident. As the name implies, the immediate causes of an injury are the causes that influence the accident just before it happens. Taken together, the immediate causes listed must be sufficient to cause the accident or injury. There will be at least one immediate cause for each of the things that came together improperly in the accident. In this case the immediate causes are:

- Eye exposed
- Sample splashed

These two causes satisfy the requirements of immediate causes in this case. Taken together they account for the accident. Separately they represent the last links in the two sides of the incident, the worker's eye and the chemical sample. The arrows in the figure indicate that they contributed to cause the outcome of *Sample splashed in eye.*

These immediate causes usually have causes, too. They are listed in sequence down the page, reaching back to the root causes of each chain of causes. During the accident investigation of this case, when the interviewer asked the worker why his eye was exposed, he said, *I did not have my eye protection on and I had my face very close to the sample.* The accident investigator therefore put *Not wearing eye protection* and *Face too close* on

Cause Tree Analysis Form

Accident / Injury : _____

Immediate
Causes :   _____   _____   _____

Secondary
Causes :   _____   _____   _____

3rd Level
Causes :   _____   _____   _____

4th Level
Causes :   _____   _____   _____

Etcetera

Figure 12-3. Cause-Tree Analysis form.

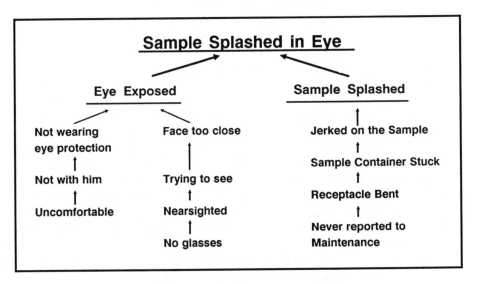

Figure 12-4. Example of Cause-Tree Analysis.

the same level below the immediate cause, *Eye Exposed.* They are written on the same level because neither is the cause of the other. They go on the level below Eye Exposed because each of these causes is necessary to account for the exposure of the worker's eye. Being a good interviewer, the accident investigator pursues each of these branches of the cause tree individually. In answer to the question of why the worker's face was too close, the worker answered, *Well, I'm nearsighted, you know, and I went off and left my glasses home this morning—so I was up close trying to see the sample.* The interviewer orders these causes from more immediate to less immediate, arriving at:

- I was *trying to see* why the sample was stuck.
- I'm *nearsighted.*
- I *didn't have my glasses* with me.

For the immediate cause of Not wearing eye protection, the interviewer elicits a similar chain of causes: not carrying it with him because he finds it uncomfortable to wear. This completes the cause tree for the *Eye Exposed* immediate cause. Pursuing the causes of the other immediate cause, *Sample Splashed,* the worker tells the interviewer:

- I *jerked the sample* container,
- Because the sample *container sticks* in its receptacle.
- This is.because the container *receptacle is bent.*
- The bent receptacle was *never reported to maintenance.*

Once again the arrows from these causes to the immediate cause indicate that they are the preceding causes and what their sequence is.

The interviewer continues the Cause Tree Analysis of each chain of causes until the point is reached where improved worker safety performance can be accurately stated in terms of objectively defined behavior. The most helpful Cause Tree Analysis is one in which the interviewer and the injured worker formulate an objective definition of safe behavior for each chain of causes. Sometimes this ideal is hard to achieve, either because of lack of time or because the immediate cause is too difficult to analyze back to its behavioral roots. In this incident for instance it might have proved too difficult to analyze why the worker's face was too close. Nevertheless, it is still important to note some general safe behavior that applies and can be defined objectively.

When the interviewer and the worker have finished the Cause Tree Analysis, they review it to be sure that it is complete. Completeness here has a very specific meaning. Cause Tree Analysis is complete when it has identified every significant chain of causes that contributed to the accident, and for each of these chains they are able to:

- specify equipment or management system changes that could prevent the injury, or to
- state the worker's improved safety performance in objective terms.

Cause Tree Analysis is incomplete if it results in suggestions for improvement that are not objectively Observable. Vague statements are not acceptable, such as:

The worker should pay better attention.
The worker should be more careful.
The worker should exercise more forethought, etc.

These are not acceptable statements because they do not specify what the worker is supposed to *do* that represents paying better attention, or being more careful, or exercising more forethought.

Pattern Search

Pattern Search is used to compare two or more incidents. A supervisor working with an employee who has had multiple incidents uses Pattern Search to look for recurring patterns of unsafe behavior in the employees work habits. In effect, the supervisor is developing an individual inventory of critical behaviors with the employee—a topic which is treated in detail in Chapter 19 on Managing the Employee with Multiple Incidents. However, whether the Planner is searching for patterns in the incidents of one employee or the incidents of an entire work group, the principles are the same. The surface particularity of accidents can obscure their underlying similarities and patterns. The Planner is looking for these underlying patterns of safety-related behavior because they are among the best clues about how to avoid future accidents. The task involves searching for common patterns of circumstance, body part, type of activity, etc. By identifying common patterns, the Planner arrives at categories such as:

- Eye and face protection
- Hand protection
- Body placement
- Tool use
- Visibility

ABC Analysis

The Steering Committee also uses ABC Analysis to uncover the existing antecedents and consequences that affect safety-related behaviors and to arrive at an Action Plan for addressing these behaviors with new antecedents and consequences. See Chapter 2 for an introductory presentation of this technique; other examples appear in Chapters 16 and 20.

THE INVENTORY OF CRITICAL BEHAVIORS

The items in a facility's inventory of critical behaviors divide into two kinds: generic and job-specific. Generic categories name the broad underlying patterns discovered in Pattern Search and other patterns generalized from them, categories such as:

- Housekeeping
- Personal protective equipment
- Body use
- Tool use
- Proper procedure

Job-specific inventory items are tied to a particular job which has been identified as a target. The procedure for developing job-specific behaviors is so similar to JSAs that it is not reviewed here. However, it is important to note that JSAs are not usually written as *operational definitions of behavior.* This means that when JSAs are used as a source for inventory development they need to be reviewed and usually upgraded to behavioral standards.

Operational Definitions of Critical Behaviors

An effective operational definition for the inventory is perhaps best understood in contrast with a poor, or non-operational definition. Remember that Observers must know whether a behavior they are watching is safe or unsafe by referring to the operational definition of the Data Sheet items. First the clear definition:

Item 4.2 Body Placement—The walking surface and working surface must be stable. Body position should not be cramped or over-extended. The worker should be balanced and have stable footing, should not be in the line of fire, or in the fall pattern of objects.

Now the unclear, non-operational definition:

The body should be kept out of way of hazards, and the worker must be aware of body placement at all times.

This is a poor definition for observation purposes because it does not spell out what it is that the worker must do to keep the body out of the way of hazards. Also, simply by watching the worker, how is the Observer to know whether the worker is aware of body placement at all times?

Job-specific operational definitions are often quite specific. The following is an example from a paper mill. A Trainer with a group of hourly employees developed this definition. It focuses on the safety-related behaviors of running a huge machine that produces large rolls and spools of paper:

Ingoing Nips—observe the area prior to starting work. Clothing and hair should be secure. Ensure balance and positive footing. Do not break down a sheet on an ingoing nip. Clean on the outgoing nip whenever possible. *When working close to nip, watch hands.* Keep hands close when possible. Stand on outside of hoses. Do not tape paper rolls while machine is running. Stop winder when cutting sheets.

Notice the italicized sentence in the definition, "When working close to nip, watch hands." This last instruction, "watch hands," sounds vague but that is not how the paper workers mean that phrase. It turns out that what they mean is literally keep the eyes on the hands when working close to the nip. This is the sign of a good operational definition. It tells the Observer what to look for. It tells the worker specifically what to do in order to work safely. The definition therefore standardizes Observer samples, and it standardizes good worker practices.

The following is one last operational definition, also from a paper mill.

Use of Paperknives, Blades, Spears and Pike Poles—Make sure that blade is sharp. Ensure good balance, positive footing. Cut at an angle when you make contact with the spool. Do not break tips off blades—change blades. Be sure hand is positioned behind blade. Keep arm up off of paper roll. Cut away from body using the dominant hand. Cut above or below the last line of cut. Start cutting 1 in. from the end of the roll. Do not use knife as a pry-bar. Do not slash. Do not stand in front of doctor blades when closing up. Put knife away as soon as job is completed.

Observation Data Sheet

The end product of inventory development is two documents, the Data Sheet and the set of operational definitions. The definitions provide a standard of safe performance; they state what to look for, and they increase the consistency and reliability of the observations. The Data Sheet is treated at greater length in Chapter 14. Figure 12-5 shows a basic Data Sheet. The salient points to note here are that:

1. The only identification that goes on the completed sheet is the Observer's. The Observer does not write down the names of observed workers.
2. The generic items do not overlap each other. Each one is mutually exclusive so that the Observer has only one place to score a given

---

Observation Data Sheet

Location _____ Observer # _____ Date ____

1.0 Procedure	Safe	Unsafe
1.1 Confined space	_____	_____
1.2 Breaking-into process	_____	_____
1.3 Pre-job inspection	_____	_____
2.0 Protective Equipment		
2.1 Eye and face protection	_____	_____
2.2 Hand protection	_____	_____
2.3 Foot protection	_____	_____
3.0 Tools		
3.1 Condition	_____	_____
3.2 Use	_____	_____
4.0 Body Use		
4.1 Visibility	_____	_____
4.2 Body placement	_____	_____
4.3 Line-of-fire	_____	_____

Category / Item   Comments

_____

_____

_____

_____

Figure 12-5. Observation data sheet.

behavior. Otherwise, there is too much room for Observers to interpret what they see, and the data will be ambiguous.
3. The Data Sheet is compact and user friendly.

These points address three of the four most typical pitfalls confronting the
Steering Committee developing a facility's inventory of critcal behaviors.
They sometimes try to do too much on the first approach and instead of
achieving point 3, the result is an unwieldy Data Sheet that is very hard to use
because of its length and detail. Sometimes the Data Sheet is flawed by
overlapping items, cases where the Observer has two, or more, places to
register the same observation. This is not good and violates point 2. It is
sometimes necessary to provide training to managers to be sure that they do
not use the observation data for punitive or disciplinary purposes. Even to

imply such a thing is in violation of point 1. To employ observation data for disciplinary purposes is very counterproductive. Nothing compromises the behavior-based approach faster than the perception that it is just a new and better way of writing tickets on people.

The fourth important pitfall is simply failing to develop one's own inventory. It bears repeating that the behavior-based continuous improvement process requires adaptation, not adoption.

SUMMARY

The facility inventory of critical safety-related behaviors is one of the core products of the Implementation effort. Given its importance not only for Implementation itself, but for the long-term objective of establishing at the facility the self-regulating safety mechanism, it makes good sense not to rush or take shortcuts. Having thus far avoided the four pitfalls listed, the Steering Committee next checks its work in the most effective way possible. The preliminary behavioral inventory and Data Sheet for the facility are presented in Inventory Review Meetings that get plant-wide feedback and input. These meetings offer numerous opportunities for furthering the Implementation effort. Chapter 13 offers a brief presentation of the Inventory Review Meeting.

Chapter 13

Inventory Review Meetings

This chapter concerns a transitional step of Implementation, Inventory Review. At this point in the Implementation effort, the Steering Committee has produced a preliminary inventory, complete with operational definitions and a facility Data Sheet. Workers have been recruited for Observer training, during which they will learn how to use the facility Data Sheet and its operational definitions. It is at this juncture that the inventory needs to be reviewed and endorsed. Facilitating these meetings requires some of the skills listed in Figure 13-1. Effectively conducted Inventory Review Meetings offer valuable opportunities for employee involvement and they build the momentum of the Implementation effort. These meetings also afford another opportunity to include important units of the facility who may not have had direct representation on the Steering Committee, etc.

On the other hand, this step also offers a forum for people to exercise their resistance to change, and to try to slow the Implementation effort with questions, and with proposed refinements of the inventory. For all of these reasons these meetings need to be conducted by Facilitators who are well versed in the inventory itself and who are skilled in managing resistance to change and other communication techniques.

Preceding the Inventory Review Meeting

By this stage of Implementation the Steering Committee has:

- Developed the initial set of operational definitions for the identified critical behaviors and
- designed the observation Data Sheets.

The Committee is doing work in progress on:

- Determining the basic unit of Observation/Sampling,
- choosing Observers to perform the baseline measurement of the facility's safety performance,
- designing and printing the feedback charts, and
- selecting the site locations for posting the critical behaviors feedback charts.

FACILITATING
MEETINGS

Skills :

Analysis
- Interpreting
  Observation Data

Communication
- Conducting Group
  Problem-Solving
- Presenting Observation
  Data
- Giving Effective Verbal
  Feedback
- Managing Resistance to
  Change

Products :

- Safety meetings
  integrated into the
  behavior-based
  process

Figure 13-1. Facilitating meetings.

The Focus of the Meeting

Given the character of the above steps, the focus of the Review Meetings is
to:

- Introduce the Steering Committee's work to key people, including hourly
  workers, supervisors, and managers,
- check their understanding of the process, and get their evaluation of the
  inventory of critical behaviors and its operational definitions before
  they are field tested,
- secure their endorsement of the inventory of critical behaviors, and
- give the inventory wide exposure

The overall goal is to manage the development of the inventory of critical
behaviors with good public relations and a spirit of participation of key

players—in short to give key personnel a sense of ownership of the behavior-based process. The more people who wholeheartedly endorse the facility's inventory of critical behaviors, the better. Certain groups are more essential than others, however.

Observers. Since the Observers will use the inventory and the Data Sheet to do their observations, they must understand and endorse both of these instruments. These people will have special concerns, and this is the time to address their questions and the issues that they raise. These issues are discussed later in this chapter.

Supervisors and managers. If supervisors and managers are to effectively support the safety process, they must understand that the inventory truly represents behaviors that are critical to safety performance in their areas. The supervisors and managers will also have their special concerns and questions, many of which are related to the issues raised by the Observers. For this reason it is usually good to combine these groups in the same Meetings.

Hourly employees. Also, since hourly employees are the people who perform the behaviors in the inventory, it is very important that they endorse it and the Data Sheet. They need to be included in the Review stage.

CONDUCTING THE INVENTORY REVIEW MEETING

At the Inventory Review Meetings each participant receives a copy of the preliminary inventory of critical behaviors, the operational definitions of the items in the inventory, and the observation Data Sheet. The Meeting Facilitator explains that these instruments were developed by fellow employees as an inventory of behaviors that are critical to safety performance. The Facilitator describes how the inventory, the definitions, and the Data Sheet were developed, and how they fit into the behavior-based accident prevention process. This part of the presentation involves reviewing the basic concepts of the process itself and draws upon the materials in the Chapter 2. The main points for presentation are:

- Accidents are a result of behavior.
- "Behavior" means an observable act.
- Since safe behavior can be observed, it can be measured.
- Because behavior can be measured, safety can be managed.
- Managing safety requires that one do the following:
  1. Identify behaviors that are critical to safety
  2. Define these behaviors in operational terms
  3. Train Observers to track the frequency of these behaviors, and
  4. Provide safety performance feedback to the workforce.

With these things in mind, and in preparation for a discussion, the Facilitator recounts the history of the development of the facility's inventory of critical behaviors. The important background facts to cover are such matters as:

- Who was involved in developing the facility's inventory of critical behaviors.
- How they went about their task—for example, the number of accidents they reviewed.
- How long a time they spent at the task.
- What percent of the facility's accidents are accounted for by each of the items or categories in the inventory of critical behaviors.

Discussion

After the group understands the background of the facility's inventory of critical behaviors, the next step is to deepen their commitment to it. This is achieved by discussing the individual items or categories of the inventory. The group is encouraged to read through the inventory items with the primary question in mind:

Will we be safer if the critical behaviors are performed correctly?

If the meeting is a large one, or if the facility's inventory of critical behaviors is lengthy, it can be helpful to divide the group into smaller units for this exercise—assigning just one category of inventory items to each small group, for instance. The participants read the inventory item/s, and group discussion follows. The goal of this discussion is to develop a consensus that the facility inventory does indeed contain the behaviors that are critical and that performing them safely will certainly increase the safety of the facility. During the course of this discussion it is important to avoid such common sidetracks as the following.

The facility's inventory is not perfect. Later in the meeting the participants will have a chance to offer suggestions for how to improve the facility's inventory of critical behaviors. At this stage of the meeting, the important point is that a facility's inventory does not have to be perfect. In fact, it may never be perfect. Perfection is not the issue. At this stage of the review, the only issue is the one formulated in the question: Will it be safer if the critical behaviors are performed correctly?

The critical behaviors in the inventory are nothing new. This will probably be true. In facilities where it is true, the important point to stress is that usually people do not get hurt in new ways but in the same old ways. The point is that the facility's accidents indicate that these small, familiar behaviors are the things to be remedied.

What is necessary to get people to perform these behaviors correctly? This is a good question, but not at this point. The behavioral safety process as a whole provides the tools to help people make continuous improvement in the critical behaviors in the inventory. Overnight change is not expected. A gradual, but noticeable, increase in the percentage of correct performance of the targeted behaviors is expected. At regular intervals, performance feedback will be posted in the work place. Observers, Safety representatives, and supervisors will interpret the charts to their crews, stressing the improvements in performance and then indicating areas for further improvement. The principles of behavioral science, show that such timely, consistent, positive consequences are the most powerful reinforcements for change in our behavior.

Therefore, all of the foregoing being true, the primary question before the group at this point is, *Will we be safer if the critical behaviors are performed correctly?* In addition, at this time it is advantageous to get the views of a wide variety of employees on how to improve the Steering Committee's proposed inventory.

Evaluating the Inventory, Definitions, and Data Sheet

After the discussion of the facility's inventory of critical behaviors, the group is asked for input on how to improve it. Relevant suggestions will address such crucial issues as:

New items. Does the inventory need to include additional items?

No overlap. An effective inventory of critical behaviors has items or categories which do not overlap each other. A Data Sheet with overlapping items is hard to fill out consistently because it has two or more places to log one and the same observation.

Observable actions only. The items of a functional inventory address observable actions only, not attitudes.

Operational definitions only. Functional definitions are strictly operational, telling the Observer which behaviors to look for and whether they are safe or unsafe.

User friendly. The observation Data Sheet should be clearly organized and user friendly.

The purpose of this discussion is input which can be used to modify and improve the inventory of critical behaviors. At issue for the evaluation are such questions as:

- Does the group see areas of the inventory of critical behaviors that could lead to problems?
- Do the Inventory, Definitions, and Data Sheet look workable?

The considered and judicious assessment of the group is what the review process needs at this stage, so that the input from the Review Meeting can be conveyed back to the Steering Committee for their consideration. The Meeting Facilitator makes good notes of these suggestions and follows through, reporting them back to the Development Committee.

Addressing the Issues

The employees who attend the meeting may have questions such as:

- Will Observer data by used against people?
- Will the supervisor really arrange coverage for the person who is doing an observation.
- Are we expected to correct each other?
- Who will the Observers be?
- What is the advantage of being an Observer?

In preparation for the Inventory Review Meetings, the Steering Committee considers these issues and draws up a set of answers to them. In order to make sure that the workforce receives consistent information, all of the Meeting Facilitators are instructed in these answers. For a detailed presentation of these and related issues, see the Chapter 14.

SUMMARY

The Inventory Review Meetings make a very important contribution to the Implementation effort. For this reason the effective Steering Committee is careful to prepare for the Meetings. Review Meeting participants need to have a sense that the Implementation effort is practical and credible at this stage, and the best way to achieve that credibility with the workforce is for the Steering Committee to anticipate and address as many of the questions and issues as they can. Of course, unanticipated questions will arise—that is the purpose of the review, to discover unanticipated issues. The Steering Committee demonstrates competence by anticipating and addressing many of the facility's questions. The Committee demonstrates flexibility and responsiveness by incorporating beneficial suggestions concerning the facility inventory and Data Sheet. This fine-tuning of the behavioral inventory and the Data Sheet make the Observers' work easier. The Observers are new to their task, but they know that the Data Sheet they are using has undergone review by the workforce. This fact assures them of a smoother reception while they are doing their observations, and it means that every effort has been made to make the Data Sheet "user friendly." These are important concerns of most Observers during their training. Chapter 14 takes up the subject of Observer training.

Chapter 14

Training Observers to Measure the %Safe and Provide Feedback

THE ROLE OF THE OBSERVER

The Observer or Sampler is a key player in behavior-based accident prevention, see Figure 14-1. As their name implies, Observers make the regular observations which provide measurement and immediate feedback about safety performance as well. These observations are also the basis for ongoing safety problem-solving and continuous improvement. In addition the Observer is the champion for the accident prevention process itself. This second function is very important. Credibility in the organization, especially with peers, is essential in an effective Observer. For reasons of employee involvement, a high proportion of Observers—at least 50%—should be hourly employees. Managers can also be especially effective advocates of the safety process when they have had Observer training. It gives them hands-on experience with the process and puts them in a position to understand fully what the inventory and Data Sheet are like to work with. They learn that behavior-based observation is harder than it sounds at first. There is room for mutual respect here.

The ideal Observer is a person who:

- Has high credibility with peers
- Is knowledgeable about the work to be observed
- Has good verbal and interpersonal skills.

The effect of these traits is that people will listen. This is a very important factor. As was presented in Chapter 2, the most effective way to change a behavior is to change its consequences. One of the most powerful consequences is information or feedback about performance, especially since workers are often unaware of the ways in which they expose themselves to injury.

PURPOSES OF OBSERVATION

The two main purposes of observation are regular sampling of the safety process, and feedback, primarily to individual workers. Injuries are the product of a system, a complicated human behavioral system with elements

```
┌─────────────────────────────────┐
│                                 │
│     OBSERVATION                 │
│                                 │
│     Skills :                    │
│       Analysis                  │
│       ● Behavioral              │
│          Observation            │
│                                 │
│       Communication             │
│       ● Giving Effective        │
│          Verbal Feedback        │
│       ● Managing Resistance     │
│          to Change              │
│                                 │
│                                 │
│     Products :                  │
│       ● Completed Data          │
│          Sheets with % Safe     │
│          ratings and special    │
│          Comments               │
│       ● Input on new            │
│          behavioral inventory   │
│          Items                  │
│                                 │
│                                 │
└─────────────────────────────────┘
```

Figure 14-1. Observation.

such as equipment maintenance, production pressures, safety training, etc. When injuries occur, the system is out of control. When a machine is producing a defective product, it needs adjustment. One way of discovering whether a production process is out of control is to sample the product. When it is possible, however, a better way is to sample upstream process indicators, such as temperature or pressure. Steering by reliable upstream indicators allows adjustment of the process before defective product is made. Similarly, a system that is producing injuries needs to be adjusted. However, adjusting the system only in response to injuries introduces an unnecessary delay. When the system is out of control, this fact is first shown by high levels of unsafe behavior. Levels or frequencies of unsafe behavior are leading indicators of injuries.

This unsafe behavior may be the kind which directly exposes a worker to injury—improper lifting. Or it may be behavior which indirectly exposes other workers to injury—the mechanic who fails to reinstall a safety guard removed while fixing a machine. In the first case, the %Safe rates can be measured for lifting by observing numbers of workers who are lifting things. In the second case, the mechanic might be directly observed to walk away from the machine without replacing the safety guard, or the Observer might later note the "footprints" of the mechanic's behavior by observing that the repaired machine was missing a safety guard. Systematic observation of safe and unsafe behaviors is a way of monitoring whether a facility's safety system needs adjustment because it has begun to go out of control.

The other main reason for doing systematic observations is to provide feedback to individuals. Injuries often occur when people are doing jobs that they do routinely and do unsafely. Workers are often unaware that they are doing a job unsafely; their unsafe routine has become a habit. A systematic observation procedure ensures that workers regularly receive information from an Observer about their safety-related behaviors. Since this information emphasizes the positive aspects of safety performance by consistently noting areas of improvement, Observer feedback becomes a soon-certain-positive consequence for safe behavior.

Although it might seem that a formal observation system is not necessary, that a foreman could give feedback to workers in the normal course of events, the problem is that in the normal course of events there are very few natural consequences that support and maintain this kind of foreman behavior. When foremen are asked how often they say something to workers about safety during the course of a shift, most of them estimate that they say something once every week or two weeks. This is not nearly often enough to change unsafe behaviors that have become habitual. Furthermore, most foremen admit that when they do talk about safety with a crew it is almost always to say something negative, not something positive. In other words, in the normal course of events a foreman's remarks provide a consequence for safety that is infrequent, uncertain, and negative—the weakest kind of consequence there is.

Observation Procedures and Schedules

Many companies have some kind of safety inspection program. These programs are useful in identifying certain kinds of hazards, but there are three factors which limit their utility in reducing injuries: focus, frequency, and thoroughness. Most safety inspection programs focus on facilities, equipment, and housekeeping, because the goal of the inspection is to identify static hazards—things that need repair, replacement, or cleaning up. These are important problems to correct; however, the typical safety inspection ignores the issue involved in most injuries—behavior.

A second problem with safety or housekeeping inspections is that they are done relatively infrequently. Monthly inspections are unusual, most of them being done quarterly or annually. This inspection schedule is probably warranted in the case of facility conditions and equipment. These things change fairly slowly, and repairs and modifications of these also take time. Consequently, more frequent inspections might not be worthwhile. Housekeeping, however, is a different matter. It is quite common to see workers scurrying around before a housekeeping inspection, cleaning up their area so that it will look good for the inspector. Poor housekeeping often represents a safety hazard, and therefore the work area needs to be maintained in an orderly state at all times, not just before a quarterly inspection. Infrequent housekeeping inspections can do little to encourage better routine housekeeping. High standards of routine housekeeping are a product of individual behavior, and individual behavior is unlikely to change with only quarterly feedback. The same holds true of infrequent safety inspections.

Finally, many existing programs fail to realize their potential because the inspections are not sufficiently thorough or rigorous. The most common type of inspection procedure consists of a group of people unsystematically looking around in an area to see what they can see. Hazards are bound to be overlooked with this kind of approach, especially unsafe behaviors.

Without a systematic approach many critical events are simply missed. An effective system of direct inspection needs to be focused, frequent, and thorough. This is all the more true when the goal is the observation of safety-related behaviors. As part of a systemic approach Observers need to be trained, something that is rare in most inspection programs.

OBSERVER TRAINING

After the Inventory Review Meetings have concluded, and the Steering Committee has had a chance to incorporate any improvements into the inventory and Data Sheet, Observer Training sessions begin. The purpose of these sessions is to provide knowledge, skill, and practice in basic observation techniques to Observers, supervisors, and other managers. The five basic skills of Observations are:

1. *How to see* safe and unsafe behaviors.
2. *How to record* what they observe—the scoring procedures.
3. *How to calculate* %Safe.
4. *How to chart* %Safe.
5. *How to provide feedback* on what they observe.

What is Behavior-based Observation?

Behavior-based Observation requires registering what is going on in the work place and judging it as either safe or unsafe on the basis of the facility inventory. When observations are done in a standardized, systematic, scientific way, they provide a measure of work place safety. One observation by itself is a sample of work place safety. The accumulation of these samples begins to develop a reliable picture of the facility's safety as a whole. High quality observation done over a period of time sketches a trend, a picture of how the facility is changing in time, either growing safer or less safe. This trend phenomenon is very important. The trend toward unsafe performance is a warning to the facility that it is asking for an accident to happen. On the other hand, the trend may be toward higher and higher levels of safe performance.

Obstacles to Observation

Behavior-based Observation takes time to learn, and there are a number of obstacles to doing it well.

Over-familiarity with the work. An Observer who knows the work too well may be complacent about the way that co-workers are doing it. In effect, in this case the Observer trusts habit more than the Data Sheet. In this respect, observation is a bit like being a pilot who is flying by instruments. The pilot learns to trust them and respond accordingly.

Unfamiliarity with the work. On the other hand, if the Observers are not familiar with the work they are Observing, they do not know what is going on. They are faced with additional work—they must grasp the situation, not just recognize hazards that they already understand.

Unfamiliarity with the facility's Data Sheet. Another problem that Observers have is that they spend less time looking at the work than at the Data Sheet. This is a typical problem for new Observers. Thorough familiarity with the Data Sheet and with the list of definitions of critical behaviors cures this difficulty.

Behaviors happen fast. It does not take very long for a worker to bend over and grab something from the floor. Did she do it properly? Before the driver changed lanes, did he look over his shoulder as he should have, or did he just glance in the side rearview mirror? Little things, and the absence of little things, count. There is really no time to notice unless the Observer has become very attuned to safety issues.

Little things add up. This problem is compounded by the fact that when things go wrong they can go from safe to unsafe instantly. The importance of little things is magnified in a crisis. A door that is half open may not appear to be a hazard, but when some unexpected thing happens, that door can become a serious danger. People have been injured walking headlong into the edge of such doors when the lights went out. A wheelbarrow in an aisle may not amount to much of a hazard, until there is a fire and that escape route is obstructed. Not wearing a seat belt can seem like a little thing; but on average, approximately once every fifty years of driving it may suddenly become a matter of life and death. Not cleaning up a spill right away can seem like a little thing that does no harm; but the law of averages is at work here too, and that harmless little spill of water or oil on the floor can suddenly become a critical contributing factor in an accident.

Two Kinds of Observation

To counteract the obstacles to observation there are two strategies, both of which require a trained eye. They are Situation-Centered Observation and Data Sheet-Centered Observation.

Situation-Centered Observation. In situation-centered observation the situation itself guides Observers. Standing back, taking their time, they let the situation show itself, as though they were seeing it for the first time. The primary question for the Observer is, *What is the potential for injury here?* This type of observation is a good antidote to being over-familiar with a work situation. The Observers see things they never noticed before. Situation-centered Observation requires real discipline. Inexperienced Observers tend to skip over this kind of observation because it can be frustrating and can seem unproductive. They find it difficult to really look without knowing quite what they are looking for. The clue lies in the question, *What is the potential for injury here?* The operative word in the question is *potential.* In behavior-based safety management, potential does not mean maybe. The injury potential of a situation does not mean the people *might* get hurt there. The injury potential is more urgent than that. It refers to how people *will* get hurt, given enough time and the "right" conditions. The potential exists at the moment of observation. It is there for the skilled Observer to see.

Data Sheet-Centered Observation. In Data Sheet-centered observation, the Data Sheet is used like a check list, ensuring thoroughness of observation. This type of observation is easier than situation-centered observation. Nonetheless, in order to be truly accurate at it, Observers need to know the Data Sheet from memory. Otherwise they are looking at it and not at the workers.

The Seven-Step Observation Procedure

The goal of the seven-step observation procedure is standardization and thoroughness. It is important that all Observers do their observations in the same way. And thoroughness is important because the Observations need to cover *all* of the same ground. Thoroughness is achieved by having the Observer do both situation-centered and Data Sheet-centered observation in one procedure.

1. Go to the action. This means doing the observation where things are happening—the Observer looks for action.

2. Look at people as much as possible. This does not mean that the Observers should not look at things and conditions. When they look at them, however, they must consider what the conditions indicate about the *behavior* of people. The way the boxes are stacked over there, is it a sign that someone has moved them by hand or with a lift? This is the kind of question the Observer asks continually.

3. Introduce yourself. When Observers begin, they introduce themselves to the workers and explain what they are doing. They are not spies, and they show people their Data Sheet and talk with them about the observation process. Observers are a champion of the process, telling people to continue with their work and that they will be told what was Observed when the Observation is finished. If the workers express concern about being observed, they are assured that no names are logged and that no disciplinary action will result from the observation.

4. Situation-Centered Observation. The Observer takes time and studies the situation, looking for potential injuries. Effective Observers do not go on to the next step until they either have a sense of potential injuries in the situation, or see that the situation is fundamentally safe.

5. Data Sheet-Centered Observation. Now the Observer goes down the Data Sheet like a check list, very systematically.

6. Give verbal feedback. After the Observer has logged the safes and unsafes and has calculated the %Safe figure, he or she is ready to give feedback on what was observed.

7. From start to finish—20 to 30 minutes. The whole procedure, including calculations and feedback, should take only 20 to 30 minutes.

Verbal Feedback—Tips for Talking

Observers provide verbal feedback and discussion following an Observation. This amounts to talking with the employees observed about what they have seen and noted on the critical behaviors Data Sheet, and why they noted what they did. The technique for providing this feedback follows a proven sequence. Positive feedback is given first. The Observer talks with the employees about the safe things he saw, emphasizing especially those things that demonstrate improvement over previous observations. The Observer then talks about areas that need improvement. Their manner is helpful throughout, making suggestions, asking questions, encouraging questions from the employees, and actively engaging in problem-solving with the workers observed.

The following are some Tips for Talking that Observers practice during Observer Training:

Prevent the accident. Observers who see that someone is about to get hurt stop the accident from happening.

Respect the people who are being Observed. They know what they are doing, and they probably have reasons for doing the job the way they are. It is not the Observer's job to boss them. The Observer and the workforce share a common ground—no one wants an accident.

Stick to the facts. When Observers are discussing behavior, they stick to the facts and do not talk about people or preach to them about safety.

Be specific. The Observer cites specific things so that people know what the feedback means.

Acknowledge people's progress. The Observer emphasizes improved performance as well as discussing areas for further improvement.

Discuss and ask. When something that workers are doing looks unsafe to the Observer, the Observer discusses it with the workers and asks questions about the situation. In such a discussion the Observer is engaged in the first step of ABC Analysis—the aim is to determine what antecedents are triggering the unsafe behavior, and what consequences are reinforcing it.

Do not argue. The Observer does not argue with someone who is resistant to the observation process.

The Scoring Procedure—Using the Data Sheet

For a presentation of a sample Data Sheet, see Figure 12-5 in Chapter 12. Use of the Data Sheet proceeds in two stages, scoring and calculating. The scoring procedure has four principal rules.

1. Do not score what you do not see. This is a reminder to the Observers not to "interpret" what they see.

2. Any safes for an item—score 1 safe for that item. For instance, suppose the Observer is doing an observation of a job for which gloves are required by the facility behavioral inventory. As the observation begins the Observer notes that all fifteen workers in the area are wearing gloves. Not seeing any unsafe behavior for this item, the Observer gives the relevant item (Hand protection) on the Data Sheet a score of 1 safe. As long as all of the workers present are behaving safely, their safe behavior gets for a score of 1 safe for the relevant item on the Data Sheet.

3. Any unsafes for an item—score 0 safes for that item (unsafe cancels safe). However, suppose that during the observation of the area three workers arrive and begin to work there, two of whom do not put on their gloves. There are now eighteen workers present—sixteen with gloves and two without. The unsafe behavior of the two workers cancels the 1 safe score arrived at above, resetting it to zero.

4. Add up the unsafes for an item. For the two workers who were not wearing gloves in an area where gloves are required the Observer would then enter a score of 2 unsafes for the item, Hand protection.

Summarizing the rules, there are only two possible scores in the Safes column, 1 or 0. The Unsafes column on the other hand may range from 0 through as many particular unsafe behaviors as are observed. For instance, if of the eighteen workers observed above, only two had been wearing their gloves and sixteen had been without them, the score for the item would have been 0 Safe and 16 Unsafe.

The result for the safety effort is that these four rules provide a weighted scoring system that is very sensitive to even slight changes in the frequency of unsafe behaviors. The Observer then totals the weighted scores from the various items of the Data Sheet and, using a formula, calculates the %Safe figure for the observation. This result is plotted on the work group feedback chart, and the Data Sheet is posted by the chart. The work group has a chance to review the observation, and then the Data Sheet, or a copy of it, is routed to the data management personnel. (For more on data management, see Chapters 9 and 10.)

Other Contributions of the Observer

Initiating job actions. It may be the Observer's job to initiate action on safety-related maintenance items. Even in cases where it is the supervisor's responsibility to initiate such action, the Observer's role is to make sure that the information is presented clearly to whoever will take action. Often the Observer also has the responsibility of following up the action.

Safety meeting resource. The Observer has an important role at safety meetings when the behavioral data is analyzed by the group for the purpose of problem-solving. The Observer amplifies the information contained in the summary data sheet reports, providing general impressions and giving the benefit of his or her unique perspective based on experience with the observations in question.

No spying. The effective Observer is not a "safety cop" or authoritarian figure of any kind. The Observer does not sneak around trying to catch people doing something wrong. The Observer does not report the names of observed employees to anyone for any reason. Nor is the Observer expected to force behavioral change on the people observed. The Observer or sampler is there to provide a measurement of safety performance, to make suggestions for improvement as well as to recognize improvement with feedback that is soon, certain, and positive.

Rotation. It is a good idea to change Observers periodically. However, care should be taken not to do this too frequently. Six to twelve months is a good period for an Observer. Initially it is a good practice to give Observer training to all supervisors and managers, up to and including site manager, and to about 10% of the hourly employees. Subsequently, additional groups of employees are trained, with the long-range objective of training all employ-ees in observation skills. Rotating Observers periodically provides an ongo-ing source of renewal for the accident prevention process and the benefit of different points of view. There is also the fact that people who have been trained as Observers become more sensitive to their own behavior. For this reason companies often train an entire workforce to be Observers even though only 10% of the employees function as designated Observers at any given time. It can be very helpful for the workforce to know from the outset that eventually everyone will be an Observer. This expectation makes the work of the first group of Observers easier, and it makes it easier for the other workers to accept the idea of being observed.

Necessary Skills and Knowledge

- Foundation Concepts
- Observation Techniques
  - Seeing Behavior *versus* Conditions
  - Familiarity with the Data Sheet and Definitions
  - Consistency of observation procedure
- Feedback Techniques

Foundation concepts. Since Observers are champions for the safety process, especially during the early phases of implementation, it is very important that the Observers understand the rationale behind the behavioral approach. An Observer needs to know why the approach works the way that it does, what the basic concepts are, and how these concepts are translated into action through the behavioral process. It is essential that the Observer be familiar with the material contained in Chapter 2.

Observation Techniques—seeing behavior versus conditions. Behavioral observation is an acquired skill. Experienced Observers understand what the basic issues are in being able to *see* behavior. They have developed the ability to look with discrimination at an activity and to see the aspects of it that are representative of the behaviors targeted on the facility's Data Sheet. The inexperienced Observer, whether hourly employee or plant manager, finds it very difficult to focus on behaviors. This is because the inexperienced Observer is much more inclined to look at the facility and its condition rather than at the actions of the employees.

This inclination is "natural." There are a number of reasons that an untrained or inexperienced Observer tends to register things rather than behaviors. For one thing, behavior happens fast. Oftentimes critical safety-related behavior happens very fast—like a play in basketball or football—and this makes it hard to observe this behavior with certainty. Compared to the confidence Observers feel in reporting on the physical plant, their confidence in themselves as Observers of behavior can be quite low to start with. For example, was the observed employee really standing in the line-of-fire? This behavior may occur for only a brief moment of time, and yet it is very significant. Is the observed employee lifting properly? The duration of the actual lift may be very short but nonetheless of critical importance to the safety of the employee. Skilled Observers learn to *see* the critical behaviors, to have *confidence in their ability to record* the behavior they have seen, to *convey* with appropriate feedback what they have seen.

Observation Techniques—familiarity with the Data Sheet. Experienced, well trained Observers produce an accurate measure of safety performance because they have achieved fluency in their use of the Data Sheet. They are well acquainted with operational definitions for each of the behavioral items on their facility's data sheet, and they observe for *each* of these items during *every* observation that they conduct. The Observers who have not yet learned their Data Sheet have a tendency to be distracted and to fall back to looking at conditions. They become side-tracked by various issues other than behavior, and the result is a measure of decreased reliability.

Observation Techniques—consistency of Observation procedure. Experienced and reliable Observers are knowledgeable about the proper steps of the observation, careful to perform *all* of the steps and *in sequence.* This carefulness extends from the larger issues of focusing on behavior and familiarity with the data sheet to such "smaller" procedural questions such as where to obtain fresh copies of the Data Sheet, how to mark them, and where to submit them for compilation. In addition the Observers should be clear about the number of observations they are to conduct and at what intervals. Skilled Observers make it their responsibility to observe at the desired frequency, generally twice per week. Carried out by 10% of the hourly employees performing observations, this schedule yields about one observation per day per work group. The observation schedule is fulfilled even when the work group is busy on a special project.

Feedback Techniques. The critical point here is that the Observer know how to talk to other employees so that:

- they listen and join the discussion, and
- the discussion is productive of improved safety performance.

Training Strategies

In addressing the training needs of Observers, several approaches are possible —each of them has its pluses and minuses.

Individual Self-Instruction. Individual self-instruction provides the Observer trainee with a set of training materials—manuals, workbooks, and video tapes—arranged in a series of presentations. The advantage of this approach is that it can be self-paced, and therefore meet the needs of a broad range of individuals. There are several disadvantages to this approach. It is dependent on the individual trainee to maintain a high level of motivation. It is hard to monitor. In addition it does not develop a group spirit, and as a consequence the individual may develop misconceptions without anyone else knowing about it. There may be such inconsistencies between Observers that their observations are not reliable.

Tutorial. Another approach is the tutorial format—small groups of four to six trainees with a trainer. This training method affords each trainee the opportunity to ask many questions and to go into the material in depth. Practicing new skills in small groups rapidly builds Observer confidence and increases inter-Observer consistency and reliability. The tutorial approach does not have the disadvantages of individual self-instruction. All other considerations being equal, the tutorial approach is probably the most effective.

Large group. The third method is to train all Observers as one group. This has the benefit of developing a spirit among the group which could add to the overall visibility of the effort. It also increases the likelihood of consistency of observation across the Observer pool, an important consideration. The disadvantage of this approach is that individuals may not receive the attention from the trainer that they need.

Skill Acquisition

The important functions for the Trainer here are coaching and motivation. What the trainees need most of all is a clear explanation of the material *in terms of its related skill*, an opportunity to *practice* these new skills under controlled conditions, and *motivation* to use their new skills on the job. The best sequence for the acquisition of a skill is:

- Model
- Practice
- Feedback
- Re-practice

During modeling, the trainee watches someone exercising the skill to be learned. The trainee then practices the skill in the presence of others. The others who saw the practice session then give the trainee feedback on his or her performance of the skill. Then the trainee re-practices the skill in the presence of the same group.

Such a presentation of the skill *Scoring the Data Sheet* would have the following steps. During the modeling phase, Data-Sheet scoring is discussed and demonstrated by the Trainer, using a slide presentation and data sheet forms appropriate to the facility or target area. During the practice phase, the Trainer conducts a scoring exercise presenting slides of behavior to be observed while the trainees mark the Data Sheet. Then the trainees pair up and practice an actual observation while the rest of the class watches their performance. (An intermediate step is to have the trainees watch a video presentation of critical behaviors—body position, lifting, personal protective equipment, etc.) During the feedback phase, the Trainer and the class give feedback on how the trainee/s performed the observation. Then the trainees re-practice what they have learned through their discussion of issues and compared notes.

A Typical Training Program

A typical training program in behavioral observation has the following elements:

Foundation Material—Presentation by a Trainer

Observation Procedure—How to do the observation, forms required, logistics, and expectations.

Behavioral Inventory Definitions—Slides, discussion, scoring exercises.

Verbal Feedback Techniques—Presentation, demonstration, and role-playing.

Practice—Observe, discuss, two rounds—two hours.

This training regimen is adequate to get an Observer started. In addition, the following training is used to upgrade the Observer's skill level.

Calibration and reliability—Does variation in the observation data come from the Observer or from the observed behavior? If it is the former, there is a calibration problem. The best technique for developing calibration is to have Observers make observations in pairs and then compare their rates. When their respective %Safe ratings are within 5%—10% of each other, the Observers are "reliable."

Practice improving feedback skills—In spite of training in how to provide positive feedback and how to make suggestions for improvement, Observers often remain weak in this area. They feel awkward and uncomfortable, and they shy away from one-on-one contact. The best strategy for upgrading Observer verbal feedback skill is for Observers and/or Facilitators who are skilled in feedback to accompany less skilled Observers and coach them in verbal feedback. For a detailed presentation of coaching for skills development, see Chapter 11.

SUMMARY

During the first phase of observation, the Observers charted feedback is not given to the workforce. Instead, the accumulating observation data is used to compile a measure of the facility's baseline safety performance. The presentation and interpretation of this charted feedback for the workforce is the focus of the Kickoff Meetings, so-called because they serve to initiate the behavior-based safety process throughout the target area. In fact, effective Kickoff Meetings are crucial to the success of the Implementation effort. To use an image, if the Inventory Review Meetings are a sort of dress rehearsal, the Kickoff Meetings are the grand opening. Organizing and preparing for the Kickoff Meetings is the subject of Chapter 15.

Chapter 15

Introducing the Process to the Workforce

Implementation of the behavior-based accident prevention process reaches a very important step with the Kickoff Meetings that formally launch the process in the target areas. However, this is not the first time that the workforce has encountered the process. By now the workforce has been involved in the Assessment and in the Inventory Review Meetings, and it has received various communications from the Steering Committee. By this point in the Implementation sequence Observers have been trained and have used the Data Sheets for five or six weeks to establish the facility's baseline %Safe rating. The workforce will also have some acquaintance with the training process, based on publicity designed by the Steering Committee. However, it is during the Kickoff Meeting that all of the workers as a group get their first formal introduction to the basic concepts of behavior-based accident prevention, to their facility's behavioral inventory and Data Sheet and are presented with their baseline safety performance figures and graphs. Therefore the Kickoff Meeting is an integral part of the behavioral safety process itself. Typically conducted work group by work group, the primary objective of the Meeting is to gain the enthusiastic endorsement of the safety process by the workforce.

With such important matters at stake the successful Kickoff Meeting requires thorough planning, and the Meeting Facilitator needs to be skilled, see Figure 15-1. The following discussion covers a list of things to consider and/or accomplish in preparation for the Kickoff Meeting/s.

PLANNING AND CONDUCTING THE KICKOFF MEETING

Advance Publicity. The more that people know about the behavioral safety process, the better. Articles, before and after Kickoff, in facility newsletters can be very helpful in alerting the workforce to progress in the Implementation effort. Announcements at regular safety meetings are another avenue for providing advance publicity for the Kickoff Meeting/s, as are letters and circulars.

Photographic Slides/Video Tapes. Slides or video tapes of the facility's critical behaviors can be a very helpful way to introduce the work group to the safety process. The presentation need not be elaborate or fancy; in fact, the emphasis is on clarity of demonstration. It is important that the Kickoff

```
┌─────────────────────────────┐
│                             │
│   FACILITATING              │
│   MEETINGS                  │
│                             │
│   Skills :                  │
│                             │
│   Analysis                  │
│   ● Interpreting            │
│     Observation Data        │
│                             │
│   Communication             │
│   ● Conducting Group        │
│     Problem-Solving         │
│   ● Presenting Observation  │
│     Data                    │
│   ● Giving Effective Verbal │
│     Feedback                │
│   ● Managing Resistance to  │
│     Change                  │
│                             │
│                             │
│   Products :                │
│                             │
│   ● Safety meetings         │
│     integrated into the     │
│     behavior-based          │
│     process                 │
│                             │
└─────────────────────────────┘
```

Figure 15-1. Facilitating meetings.

Meeting presenter not assume that people already know which behaviors are safe and which are unsafe.

The persons responsible for Kickoff presentations begin making and assembling the slides or video footage early, usually during the development of the inventory of critical behaviors. Slides are easier to produce than video tape is. The number of slides needed will depend on how many items are included on the facility's inventory and Data Sheet. Between 50 and 60 slides are usually adequate. This number of slides represents both the safe and the unsafe versions of the specific behaviors of the inventory, since the work group must be shown what the Observers/Samplers regard as Safe and what they score as Unsafe. If the facility Data Sheet includes conditions such as housekeeping, then the safe and unsafe versions of these need to be shown in slides also.

It is advisable to stage the scenes for the slides or videos. Recruit workers with credibility among their peers to model both the safe and the unsafe behaviors on the facility Data Sheet. This approach has several benefits. The work group sees familiar faces demonstrating the critical behaviors. The photographer sets up the scene so that it clearly illustrates main points of the Kickoff presentation. The photographer works at his or her own schedule, without having to wait for suitable opportunities to make pictures. The people in the pictures know that they are being photographed; these are not candid camera shots that might embarrass them or compromise their standing on the job.

Targeted Feedback

The Kickoff Meeting is the first presentation of observation and feedback to a work group. The important consideration is that workers must feel that the feedback is relevant to them. Especially when the baseline %Safe figures are very low, the workers need to be sure that the figures are not the fault of some other work group. Therefore the appropriate organizational level for the Kickoff Meeting is usually that of the first-line supervisor and the work group.

Since the meeting is also a time for group problem-solving and discussion to address the percentage of safe and unsafe behaviors, the format implies an organizational level where the problems are fairly common and where this kind of group process can be done effectively.

Posting Feedback Charts

Feedback charts should be posted in a very visible place, where workers will see them regularly. During their Kickoff Meeting, the workers are shown the chart for their area and are notified where it will be posted. Finding an appropriate place can present problems. For instance, in a rotating-shift environment there may be four crews who work in a given area, each of which could have a feedback chart. In which case there would need to be room to post at least four charts in one location. It is worthwhile to build or purchase display space for feedback charts. The displays do not need to be elaborate. The key issues here are the following.

Regular access. Post the feedback charts in an area where workers will see them regularly.

Good visibility. The charts should be large enough to be easily visible.

Room for Data Sheets. Ensure enough room to post observation Data Sheets along with the feedback charts.

Sheltered place. Select a place for the charts and Data Sheets that is reasonably protected from weather, dust, debris, etc.

Easy to update. Observers need to be able to get to the charts easily to post their data points.

Preparing Feedback Charts

The Facilitator prepares a chart for each work group. This may mean preparing dozens of charts. There must be a plan for providing new charts in the future. As the observations accumulate, the Observer will fill the chart. The chart for a work group has the work group name on it along with other explanatory information. Baseline data points are entered before the Kickoff Meeting.

Determining the Meeting Format

Kickoff Meetings can be done in several ways. Different organizations emphasize different things. The following considerations are relevant for any meeting, however.

Length. Kickoff Meetings usually run from two to four hours in length. A large portion of this time is used to review slides or video tapes of critical behaviors. Therefore a smaller inventory of critical behaviors means a shorter meeting.

Group size. The meetings go better if the group is relatively small. Generally, the best size group is the work group—a first-line supervisor and the people who report to the supervisor. When this is not convenient and a larger meeting is necessary, the Meeting Facilitator holds meetings later with individual work groups in order to discuss with each group its %Safe baseline levels.

Focus. If the meeting is a large one, the Facilitator has to make a judgment call about showing the slides or tapes of the inventory items. One solution is to show to the large group just the slides or tapes illustrating the inventory items that all of the work groups present at the meeting have in common with each other. Later the slides of more particular inventory items can be shown in work group settings.

Kickoff presenters—recruiting and rehearsing them.   The Steering Committee usually presents the Kickoff Meeting. It selects a variety of its members to cover the needed topics. The presenters may not be accustomed to speaking in such situations. In most cases it is worthwhile to have a rehearsal. Rehearsal gives the presenters a chance to become more comfortable with their role, and it offers an opportunity for the Meeting Facilitator to do some coaching and make sure that each presenter covers the subject matter in its proper order. Some presenters may even need written scripts or outlines at the least. It is best not to slight these preparations.

Scheduling the Kickoff Meetings.   In many organizations scheduling Kickoff Meetings can be very complex. The timing is important. For instance, it may be less effective to have the meetings immediately before extended holidays or early in the summer just before many people leave on vacation. On the other hand, it may be more important to get the process underway. In any case, an ill prepared Kickoff Meeting is always too soon.

Points to cover in the Kickoff Meeting. *Review the purpose.* The Kickoff Facilitator reviews the purpose of the meeting and discusses the agenda. Some people at the meeting will know what it is about, but an effective Facilitator spells out the purpose of the meeting.

Statement of management support. The support of the facility manager is critical to long-term success of any process. It is important that the wage employees be clear about the commitment of top management to the safety process. It is usually possible for the facility manager (or department heads) to be present at each Kickoff Meeting if the Steering Committee asks them to attend. However, if upper management cannot be present at the meetings, it is good to make a video tape in which the managers of the facility discuss the accident prevention process and clearly state their support for this process as a means of achieving a safer work place. The Steering Committee advises the manager and department heads on what topics to include in their video taped remarks. It can also be helpful to rehearse these talks so that they proceed smoothly and to the point.

In addition to a video tape, personal appearances by the facility manager and department heads at some meetings is very desirable. Furthermore, the first-line supervisor of each work group needs to be present at the Kickoff Meeting to demonstrate understanding and support of the process.

Introduce the Steering Committee. The Steering Committee not only facilitates the Kickoff Meetings, it introduces its members and its work at these meetings. The hourly employees on the Steering Committee can lend great credibility to these meetings in particular and to the process as a whole. They should be fully involved in the Kickoff Meeting preparations and in the reports and presentations made during the meetings.

Since it may not be possible for all of the Steering Committee members to be present at all of the Kickoff Meetings, they also can be introduced on video tape or in a slide presentation. It can be very helpful for the workforce to see how many people from which levels have been active on the Steering Committee. At a minimum, the function and membership of the Committee needs to be discussed, preferably by a member of the Committee.

The history of the accident prevention process at the facility. Discuss the need for a more effective safety effort at the facility, and explain how the behavior-based safety process was selected to address this need. Review the history of the behavioral safety process at the facility to date. Many of the workers will know something about it, having participated in the Assessment interviews and surveys, or having talked with the Observers while they were taking the measure of the facility's baseline safety performance. However, most of the workforce will not have a very clear idea of just how much work has already been invested in the behavioral safety process. One of the primary reasons for the Kickoff Meetings is to make it clear that this process has received considerable attention thus far, and that it has been adapted to the facility instead of being some spur of the moment, off-the-shelf program.

Introduce the Foundation concepts. These concepts need to be covered, but not in great detail. A member of the Steering Committee reviews the basic theory behind the behavior-based approach to safety management. This is usually done in a condensed summary, leaving time for a question and answer period.

Present the Observation and feedback cycle. A worker's basic question about any new initiative is, *How is this going to affect me?* The Kickoff Meeting addresses this primary question with detailed answers to questions such as:

- What is an observation?
- Who does the observations?
- How often are observations done?
- What will happen as a result of observations? (Common concerns have to do with whether there will be disciplinary action as a result of observations, and whether the Observers will be spies or snitches.)
- What kind of feedback will the Observers give?
- Who will see the observation Data Sheets and in what order?
- What is the system for tracking safety-related maintenance items noted during observations?
- What do the different graphs and reports of observation data mean?

Introduce the Observers. This is the initial, core group of Observers and usually they are very enthusiastic about the behavioral safety process. It can be very effective for them to make short statements about the process because they usually enhance the acceptance of the process by others.

Review of the Inventory. This is the heart of the Kickoff Meeting, and usually it takes the most time of any of the Meeting's elements. The essence of the behavioral safety process is letting workers know how to perform key tasks and then giving them feedback on how well they are doing. The review of the facility's inventory of critical behaviors provides a perfect opportunity to let the workforce as a whole know how to behave safely. This amounts to letting them know what the standard for safe behavior is at the facility. By showing *both* positive and negative examples of the critical behaviors in the inventory, the Steering Committee indicates to the workers what is expected and what is not wanted, thus enhancing the workers' ability to discriminate between safe and unsafe behavior. Group participation can be engaged over both the safe and the unsafe behaviors by asking people to point out the critical behaviors they see in the slides or video tapes and whether they are being performed safely. For ease of assimilation it is best to group the slides of the facility's critical behaviors, presenting all of the housekeeping items, and then all of the body-use items, for instance.

Test on critical behaviors. After showing these slides, some facilities test the participants by showing additional slides or video tapes and having people rate what they see as either safe or unsafe. The most complete version of this testing procedure involves doing a mini-training session on observation techniques, distributing a copy of the Data Sheet to each of the participants, and having them use the Data Sheet to actually score the scenes presented in the slides or video tape. This kind of testing can be very helpful. The workforce gains a better insight into the observation procedure.

Present the Observation data. Using slides, overheads, or video tapes, present work group observation data—%Safe for each Data Sheet category, for example, or overall %Safe by department or shift.

Present the feedback charts and the baseline measure. Introduce the group's feedback chart and its baseline data. Figure 15-2 shows a feedback chart at the time of the Kickoff Meetings. The baseline safety performance of Machine Shop—Shift 3, the work group charted here, hovers at 50%. (Note: if there is more than one work group in a Kickoff Meeting, this point may be postponed until the Facilitator can meet with each group separately.) Review the chart so that the workers understand how to read it and are reminded of how the chart will be updated, where it will be posted, etc. This information may already have been touched on at various points in the meeting, but it bears repeating. Let the work group know that the completed observation Data Sheets will also be posted beside the feedback chart. The Meeting Facilitator gets the group's reaction to its own baseline performance —is it "good" or is it "bad"? Does the group's data show areas where they are especially "good" or "bad"? This is a good point on which to encourage group discussion and a preview of the problem-solving techniques that will later become routine at the facility.

Invite questions and discussion of issues. Effective Facilitators are scrupulous on this point. They make time for questions and concerns, usually by announcing that there will be a time for them—the end of the meeting is a suitable time. However, as questions come up during the course of the meeting, they record them to come back to. If there are questions they

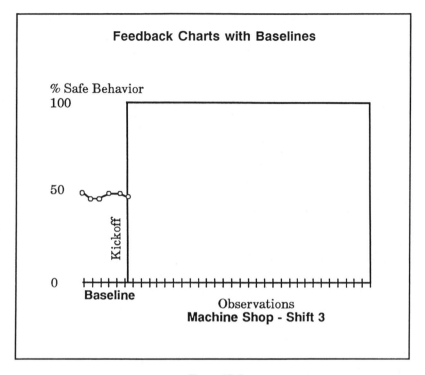

Figure 15-2.

cannot answer, the let the questioners know how and when they will have an answer.

Closing statement from the first-line supervisor. First-line supervisors play a key role in the behavioral safety process, especially after the Kickoff Meetings. While it is important to highlight the contributions and ongoing roles of hourly personnel, it is also necessary to give first-line supervisors high visibility at the Kickoff Meetings. Since the supervisor will be actively engaged in the safety mechanism as it takes root at the facility, it is appropriate to end the meeting with a statement from the first-line supervisor.

Makeup Meetings.   Due to illness, vacation time, or the need to remain at the work station, some workers are bound to be absent during the Kickoff Meetings. The Steering Committee must have a plan for Makeup Meetings.

1. The makeup session could be done in very abbreviated form, one-on-one by the first-line supervisor, for instance.

2. Or workers who missed their own Kickoff Meeting can attend with another work group—which means that they may miss some information that was particularly relevant to their own work group.
3. Or special makeup meetings can be scheduled—usually this mixes together the personnel for a variety of work groups, heightening the drawback of plan #2.

In any case, it is less important which method is used to bring people up to speed and more important that there be a plan for makeup and that it is carried out. Ignoring the need for makeup sends a negative message to the workforce about the process.

SUMMARY

Effective Kickoff Meetings set the stage for the last phase of the Implementation effort. The observation procedures gain momentum. Supervisors, Observers, and the workforce at large, learn what to expect, and they learn what is expected of them. The observation data is posted regularly for the various work groups. The work groups begin to watch their progress as reflected in the charted feedback. They note their performance marks in relation to their own past performance and in relation to the performance ratings of other work groups. In the meantime the workforce grows accustomed to the principles of the behavior-based process in the most practical way possible as the accident frequency rate declines. This situation represents a successful Implementation effort (see Chapter 17 for a range of case histories). Such a workforce is primed to play its part in the self-regulating safety mechanism. The establishment of this mechanism is the point at which the behavior-based process becomes a closed loop of continuous improvement. This safety mechanism is the subject of Chapter 16.

Chapter 16

Establishing the Behavioral Process for Continuous Improvement

INTRODUCTION

In high-functioning companies—companies with high levels of communication and cooperation—safety responsibilities may be shifted all the way to the level of the work group and its safety meetings. In companies that lack some of these conditions, the shift of roles and responsibilities to a more widely based group of employees takes the form of an intermediate mechanism. In close cooperation with the pool of Observers and experienced workers, this corps of people identifies the new areas of greatest opportunity for improvement and presents their findings to the work group for review, endorsement, and observed compliance. In some companies members of the Steering Committee specialize as facilitators, trainers, and presenters of the ongoing improvement process to the work groups as safety meetings. Figure 16-1 offers an overview of this accident prevention process.

The presentation that follows does not spell out this intermediate mechanism but focuses on work group participation instead. The rationale is that since the long-term use of this process tends toward work group responsibility for performance, that is the mechanism of the process in its complete maturity. In the meantime, however, the development of an intermediate mechanism is an effective and powerful measure. The emphasis of the description on work group practice, then, is really on broadening participation in the improvement process, with the work group representing the broadest possible form of participation.

The central point here is that the *maintenance* of the continuous improvement process has different requirements than the installation of the process. A facility with sustained improvement grows beyond the initial mechanism of the Implementation phase. As part of this more permanent footing, an intermediate mechanism may be exactly what is needed at a given facility. Where this is true, the intermediate mechanism assumes the functions that are described in the following presentation of work group participation.

Continuous Improvement is an Ongoing Process

Once the basic behavior-based accident prevention process is in place, the mechanism is used to drive the safety culture of the facility in the direction of

Figure 16-1. Data Flow in the accident prevention process.

continuous and lasting improvement of safety performance. Lasting safety improvement is a matter of cultural change. Continuous improvement is a *process*. The effective unit of this improvement mechanism is the work group that does its own safety communication and cooperates in problem identification, problem-solving, and performance tracking. As a whole, the members of the work group have endorsed both the scope and the particulars of their behavioral inventory, its operational definitions and its Observer Data Sheet. A number of them have trained and functioned as Observers, making the %Safe measures of the group. At the regular safety meetings they contribute their new skills: their ability to focus on behaviors and not be distracted by non-behavioral things and conditions, their dialogue and coaching techniques, their sense of the importance of the upstream factors of accident prevention.

At the Kickoff Meetings this work group may have included a number of skeptics, but in the meantime they have seen the Implementation effort get results. The critical behaviors targeted by the facility Steering Committee have been improving, and the associated accidents are on the decline. The work group has seen the positive and compound effect of this improvement — better attitudes, more generalization of safety-related behaviors, more awareness, better performance. Adhering to the terms established with the Steering Committee, management and supervision have been very careful to keep the observation process anonymous, with the result that the discipline issue has been dealt with satisfactorily from all points of view. Observers have been careful to follow the operational definitions that support the facility Data Sheet, writing down only what they see. The flow of feedback in charts, graphs, and safety meetings has remained focused on critical behavior, giving new life to the safety effort and stronger credibility for the idea that safety can be improved. And, finally, management and supervision have been consistent in their acknowledgement and appreciation of the improved safety performance of the work group. These conditions taken together mean that the work group is on the threshold; they are at the transition point from the Implementation phase to Continuous Improvement.

THE INTEGRATED SAFETY SYSTEM

It is at this point of demonstrated benefits that the work group closes the accident prevention loop and adopts the mechanism as its own. It may be helpful to borrow an image from automotive engineering to describe the preceding steps of Assessment and Implementation. The Steering Committee (the driver) turned the key and applied some gas. The engine was cold but it started, and now it has warmed up. At this point in the process, the engine of accident prevention (the work group) is turning over nicely. It is now to the point where the driver can ease off on the gas and the engine will supply its own needs, pumping its own fuel, generating its own current and even recharging the battery. This is the self-adjusting mechanism at work. This is the goal of the Implementation effort itself, to foster the self-adjusting work group. As safety performance ratings improve the safety culture begins to improve. This is an important development. The belief system of any culture sets the limits of what is possible. The received or inherited ideas of the safety culture are a powerful deterrent to change. The beliefs of a safety culture set the limits for safety improvement; they define what is "good enough." Non-verbally determining what is unsafe or safe, taking for granted the fact that good safety performance and high productivity are incompatible, accepting injuries as part of the cost of doing business, etc.—these are the ceilings that a reactive safety culture imposes on its members. The continuous improvement safety mechanism challenges these ceilings gradually and on the basis of objectively measurable factors.

A similar case for athletic performance. People used to believe that the four-minute mile was unbeatable. It was widely held that there was a physiological barrier in the human frame and that a man simply could not run a mile in under four minutes. Then Roger Bannister did it. Suddenly many other excellent runners found that they too could run the mile in under four minutes. What had changed? Bannister's accomplishment had removed a mental barrier to performance and replaced it with a new image to strive for. His running demonstrated that the culturally accepted belief about the four-minute mile was not necessary. Bannister's performance reopened a question that had been closed, *Just how fast can a human being run?* In a similar way, the continuous improvement safety process reopens a question about safety performance and provides an image or goal, *Can injury-free performance be reached?* What would safety activities look like (image) if this kind of performance were achieved. The answer to this question emerges in behavior-based safety management with the integration of safety systems— behavioral observation and feedback, accident investigations, safety meetings, all consistent with the quality improvement process.

The existing safety culture puts limits on human performance. Behavioral analysis of incident reports challenges a facility's ideas about what is safe and unsafe. The Steering Committee is almost always surprised to discover just which generic and job-specific behaviors actually account for their most significant safety vulnerabilities, a surprise that cuts across levels and locations. The point is clear to management representatives and to union representatives, to salaried and to hourly personnel—the existing plant-wide belief system does not really understand where the weaknesses are, let alone how to correct them.

Having identified the behaviors which make the facility most vulnerable to accident and injury, people begin to appreciate the operational definitions for what they are—an inventory of *accident-prevention* behaviors. The input of hourly employees is crucial to inventory development. Whatever they may think about safety, experienced, productive workers with good safety records know what should be done to work safely. The job of the Steering Committee during planning is to use behavioral analysis techniques to gather this valuable knowledge from the workers and then to organize and incorporate it into clear operational definitions (see Chapter 12). The Steering Committee is concerned to find out exactly what a worker should look for in pre-job inspection of the work area. Just what does safe practice mean in terms of body position and placement with respect to task? Skilled hourly employees with hands-on experience of the equipment, processes, and products of the facility do not know generic and job-specific operational definitions, but they know the details of the facility's work. Their knowledge sifted and analyzed with behavioral techniques produces the most effective inventory.

The contribution of such workers challenges the idea the productivity and safety are incompatible. Deming and others have shown that productivity follows quality. In a related way behavior-based safety shows that productivity also follows from the continuous improvement approach to safety performance. Both the immediate and the long-term results are positive. Employee involvement is sought and valued. The facility establishes a powerful new resource—a concentrated, centralized, standardized inventory of the best safety practices compiled by knowledgeable workers. From the outset the inventory of critical behaviors is designed for ready incorporation into safety meetings, training programs, special applications (back-injury prevention, managing employees with multiple incidents, etc.), accident investigation procedures, and safety performance accountability at all levels. Along the way, many unexamined "truths" go by the board, among them the belief that injuries are part of the cost of doing business.

Shifting to Wider Participation in the Process

Changing the safety culture means increasing the opportunities for employee involvement in safety assessment and problem-solving. The behavior-based safety meeting is a forum for this kind of involvement. The first items on the facility's inventory of critical behaviors were identified by the Steering Committee. There was worker representation both on the Steering Committee and during the review and endorsement process, and the Kickoff Meetings were addressed to the workforce at large. Up to this point the primary responsibility has been with the Steering Committee. The responsibility for identifying and solving problems now shifts to a wider group, with the accumulating observation data providing the framework.

The mechanism for continuous improvement—how it works. The continuous improvement safety mechanism is several things at once. It is a data-based, cooperative problem-solving process that is designed to raise the cultural standards of safe behavior. It does this by remaining evergreen, always focusing on some new area of performance once it has achieved satisfactory performance of previously identified focus areas (review Figure 2-7 in Chapter 2). The steps of the process are: review performance data, develop a focus, cooperative problem-solving, generate a behavioral Action Plan, and specify its follow-up.

At its safety meetings the work group reviews the observation data from its own %Safe charts and computer reports. For instance, see Figure 16-2. This is the same chart that appears as Figure 15-2 in the preceding chapter on Kickoff Meetings, but the chart has been updated to reflect observations of the work group in the months following the Kickoff Meeting. The chart

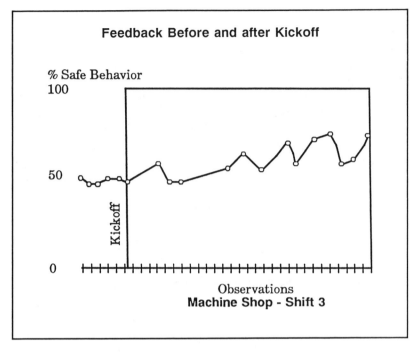

Figure 16-2. Feedback before and after Kickoff.

shows that the overall trend of work group performance is toward improvement. It also shows that progress is somewhat erratic. The work group discusses these and related matters. The work group also looks at computer reports on Observer comments and questions Observers and supervisors about their impressions of safety performance. Investigation reports of accidents and near misses are also good sources of information about areas for the group to focus on. When there have been incidents or near misses, the work group considers the following questions. Are the behaviors identified in the incidents or near misses in the inventory of critical behaviors? If the identified behaviors are not in the inventory, should they be added? If the identified behaviors are in the inventory already, are they being observed? These sources of data are reviewed with an eye to selecting a focus area, an area of special effort. The work group chooses an area that can use the additional attention, an area where higher standards of safety performance will be of significant benefit to the workers—such areas can be selected based on concerns of employees, areas of interest, etc. Likely areas are also apparent from computer reports that summarize observation data by categories or items with either a persistently low, or unstable, %Safe score, such as the category of procedure in Figure 16-3 with a low score of 63% Safe.

Machine Shop

4/10/87 - 5/10/87
Total Sheets for this Accumulation = 28
Average Observations per Sheet = 17

Summary of Categories

		Safe	Unsafe	%Safe	%Sheets
1.0	Procedure	35	21	63	11
2.0	PPE	189	14	93	79
3.0	Tools	188	14	93	46
4.0	Body Use	4	0	100	7
	Grand Total	416	49	89	

Figure 16-3. Summary report of Observations—by overall category.

Figure 16-3 presents a summary report of the main categories of critical behavior for a specified area of a particular facility as observed during a one-month period. Reading from left to right, the columns of the report present the number of safe and unsafe behaviors observed, the %Safe calculation based on those numbers, and percentage of total Data Sheets on which information about the category was reported during the observation period.

The work group could focus on a category that the Observers have been ignoring. This indicator shows up as a low percentage of observations, such as the category of Body Use in Figure 16-3, which was scored on only 7% of the Data Sheets turned in during the observation period of the report. Figure 16-4 represents a report that takes the information of Figure 16-3 and gives a more fine-grained presentation of it, showing the breakdown for the behavioral items within each category.

Machine Shop

4/10/87 - 5/10/87
page 2

		Safe	Unsafe	%Safe	%Sheets
1.0	Procedure				
	1.1 Confined Space	24	4	86	7
	1.2 Breaking-into Process	0	0	0	0
	1.3 Pre-job inspection	11	17	39	4
	Subtotal	35	21	63	11
2.0	PPE				
	2.1 Eye and Face	84	1	99	46
	2.2 Hand	42	11	79	18
	2.3 Foot	63	2	97	14
	Subtotal	189	14	93	79
3.0	Tools				
	3.1 Condition	99	2	98	29
	3.2 Use / Selection	89	12	88	18
	Subtotal	188	14	93	46
4.0	Body Use				
	4.1 Placement	1	0	100	4
	4.2 Line of Fire	2	0	100	4
	4.3 Visibility	1	0	100	4
	Subtotal	4	0	100	7

Figure 16-4. Summary report of Observations—within each category.

The work group can also turn to the Comments reports of the Observers in order to detect either developing problems, such as the Observer comment on personal protective equipment in Figure 16-5. What can be done about the fact that available gloves were not usable? Considering that vise-grips are not a "legal" tool in this facility, what measures should the work group take about the prevalence of vise-grip use? On the plus side, does the comment under Item 3.1 that *Tool condition has improved* mean that Observers can afford to sample that item less often and turn their attention to more pressing items? These are the questions that arise during work group problem-solving sessions using the observation data.

The group might also focus on a problematic aspect of the work process, such as a turnaround. They might focus on a particular employee group because of their exceptional exposure—new employees, for instance. Finally, the group might select a focus area in which they have very high %Safe scores in order to test the validity of these scores; or they might focus on some especially visible problem area so that success there will encourage employee involvement and support of the continuous improvement mechanism.

Once the group has a focus area, it uses the same tools used previously by the Steering Committee—Cause Tree Analysis of accidents or close calls, ABC Analysis of the consequences controlling a behavior, and operational definitions of generic and job-specific behaviors to be added to their inventory of observable items for measurement and tracking. Cooperative problem-

Data–Sheet Comments Report

4/10/87 - 5/10/87

		Data-Sheet #	Date	Location
1.0	Procedure			
	1.1 - Pre-job planning			
	No time.	17	4/12/87	2
	Not necessary.	32	4/19/87	1
2.0	PPE			
	2.1 - Hand			
	Available gloves not usable.	21	4/14/87	1
3.0	Tools			
	3.1 - Condition			
	Tool condition has improved.	43	4/21/87	2
	3.2 - Use			
	Vise-grips are all over the place.	49	4/30/87	1

Figure 16-5. Data sheet comments report by category and location.

solving at this stage involves brainstorming and unimpeded suggestions. The safety meeting is conducted by people who are familiar with these tools. Roughly once a year, or whenever there are major changes in the work being performed, the inventory of critical behaviors is reviewed with an eye to adding new items and their respective operational definitions. In the meantime the ongoing focus of the safety meetings is to identify behaviors already in the inventory that need more attention from the workforce and/or from the Observers.

Identifying a new item for the inventory.   When the work group identifies a target area, they do so by using many of the same planning skills that the Steering Committee used when it first developed the inventory during Implementation. The work group also brings to bear its experience in targeting areas where significant improvement is most possible. The example presented is a case in which the work group identifies the target behavior: *Wear hearing protection when required.* Observation data shows them that their performance is poor on the PPE category, and the Observer comments indicate that failure to wear hearing protection when required accounts for much of their poor rating on PPE. The work group decides to pinpoint this behavior improvement.

The work group does an analysis of *Failure to wear hearing protection when required,* arriving at the following information:

- Workers are worried about the nearby heavy equipment traffic and so do not like to have ear plugs in when they are around the compressor.
- Plugs dispenser is empty at times due to holidays, weekends, vacations, etc.
- Workers do not think that they are near the compressor long enough for the noise to hurt them.
- Not everyone remembers having been shown the hearing protection film.

ABC Analysis—Steps 1 and 2

Step 1. The first step of ABC Analysis is to state the unsafe behavior in observable terms and then to:

List existing Antecedents:

- Ear plugs are often not readily available.
- Workers have all seen the hearing protection video but many do not remember it.
- Workers are all "transient" in this area.

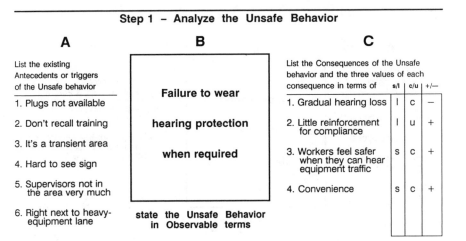

Figure 16-6. Step 1—ABC Analysis of "Failure to wear hearing protection when required."

- There is a warning sign, but it is not where it must be looked at.
- Supervisors are usually not present in the area.
- The compressor is adjacent to a heavy equipment traffic lane.

List existing Consequences and analyze for soon-certain-positive:

- Hearing damage is not immediately detectable by workers (late-certain-negative).
- There is little reinforcement for compliance with hearing protection policy in this area (late-uncertain-positive).
- Workers feel safer when they can hear the heavy equipment traffic (soon-certain-positive).
- It is more convenient not to have to hunt down ear plugs (soon-certain-positive).

Figure 16-6 shows Step 1.

Step 2. State the safe behavior in observable terms, Figure 16-7, and then:

List new Antecedents:

- Make earplugs readily available.
- Retrain employees with hearing-protection video tape—involve employees in the planning of this training.
- Move the warning sign to a more effective location.
- Construct a traffic barrier to protect pedestrians.
- Plan for hearing testing

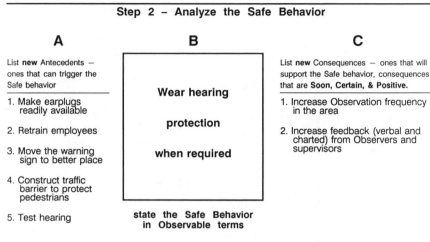

Step 2 – Analyze the Safe Behavior

A

List **new** Antecedents – ones that can trigger the Safe behavior

1. Make earplugs readily available

2. Retrain employees

3. Move the warning sign to better place

4. Construct traffic barrier to protect pedestrians

5. Test hearing

B

Wear hearing

protection

when required

state the Safe Behavior in Observable terms

C

List **new** Consequences – ones that will support the Safe behavior, consequences that are **Soon, Certain, & Positive.**

1. Increase Observation frequency in the area

2. Increase feedback (verbal and charted) from Observers and supervisors

Figure 16-7. Step 2—analyze the safe behavior.

List new Consequences:

- Increase observation frequency in the area.
- Increase feedback from Observers and supervisors.

Step 3. Drafting the Behavioral Action Plan

After the work group has developed a focus area by reviewing its performance data and by using Cause Tree and ABC Analyses wherever they are helpful, the results are formulated in a behavioral Action Plan which amplifies Step 2 by assigning responsibilities and deadlines for each item on the Action List. The behavioral Action Plan has three parts:

- Focus statement
- Action list
- Follow-up

The focus statement of the work group's Action Plan is a short, clear statement of the issue addressed by the plan, for instance, *Compliance with hearing protection policy around the compressor.* The action list details the things to be done to improve performance in the focus area—each item on the actions list specifies who is responsible for doing it and on what time table. The review measures state how the work group will know whether the focus issue has been satisfactorily resolved. The following is an example of an Action Plan.

Sample Behavioral Action Plan

Focus: Compliance with hearing protection policy around the compressor.

- The problem is that people do not always wear their ear plugs; this is especially true when they are working around the compressor.
- Behavior percentages range from 65% to 72% safe.

Patterns:

- According to Observer data, operator compliance is worse than maintenance compliance.

Actions:

1. The Medical Department will be asked to station a mobile hearing test unit at the compressor site to test everyone entering the area for one week, and to discuss the cumulative aspects of hearing damage. (antecedent intervention)
— Department Manager will arrange for this, to start in two weeks on date: _____

2. Increase Observer coverage of the compressor area for one month, to start in two weeks on date: _____; remember to provide positive feedback for improvement. (consequence intervention) — Observer

3. Instruct all Observers to discuss hearing loss with any worker who is without hearing protection in the compressor area. (consequence intervention) — Supervisor

4. Supervisors to rotate coverage of this area for one month, to start in two weeks. They will approach each person entering the compressor area with appropriate positive or negative consequences. (consequence interventions) — Department Manager

5. Supply of ear plugs to be assured starting tomorrow (date: _____). (antecedent intervention) — Supervisor

6. Operations Supervisors to show hearing loss video to their people within next two weeks and express clear and firm expectation of compliance. (antecedent intervention) — Department Manager

7. A traffic barrier will be constructed between the compressor area and the heavy equipment lane. (antecedent intervention) — To be completed in two weeks by Maintenance Department

Follow-Up:

In one month, observation data will be reviewed for compliance with hearing protection policy in the compressor area.

Accident Investigations

In the integrated safety process, investigations of accidents and near-misses proceed in behavior-based terms. The aim is to develop an investigation procedure that feeds directly into refocusing the facility's inventory of critical behaviors. This means that the investigators are not satisfied with their work until they have arrived at the critical behaviors implicated in the accident or near miss, and they can state these behaviors in operational definitions. In effect this means that the investigation procedure incorporates the tools of Cause Tree and ABC Analyses as needed.

What triggers an investigation? Which incidents require investigation differs from facility to facility, but the principle is the same: investigate any incident that may reveal something about the *potential* for accidents at the facility. This is the same thing as investigating any incident that may be serious. For instance, where the only difference between a first aid injury and a recordable is that the workers involved in the first aid incident were lucky, the principle means investigate first aid injuries. Near-misses are another important source of information about accident potential, as are accidents that destroy property but happen not to injure anyone.

Who does the investigation? The assignment of this responsibility depends on the facility. Whoever is charged with this responsibility—usually it is a supervisor—immediately initiates an investigation when an incident occurs. This expresses concern for the safety of the employees and exercises authority to assure a safe work environment. Hourly employees are involved in the investigation as an exercise of their responsibility to learn about safety. They are also integrated into the investigative procedures because of their ability to identify safety problems and to contribute to their solutions owing to their uniquely advantageous position to describe the events that led up to the incident. Sometimes outside expertise may be called for also. Thus at every level of incident investigation, the supervisor and one or more hourly employees are involved. Beyond this core of investigators the other members of the team are determined by the level of the investigation itself.

The six steps of incident investigation. The following six steps are necessary for the investigation. They differ only in how thoroughly they need to be done. The steps are:

- Fact finding, identification of critical behaviors
- Cause Tree Analysis of what happened
- Confirmation of analysis
- Researching past events
- Problem solving, and
- Devising an Action Plan

Fact finding. The person in charge of investigating the accident initiates fact finding as soon as he/she learns of an incident. For all levels of investigation, fact finding consists of interviewing the employee/s involved to discover and document what happened. Already at this stage of the investigation the effective investigator is looking for the operationally defined behaviors that contributed to the incident. These facts include: behaviors, condition of facilities and equipment, level of training, and other management responsibilities. The purpose of fact finding is not to assign blame or to arrive at some predetermined conclusion. Fact finding requires good interviewing skills, some of which are recapitulated briefly here. See Investigative Interviewing in Chapter 11 for a detailed presentation of this skill.

Investigative interviewing involves the skills of Dialogue, Listening, Cause Tree Analysis, and Pattern Search. It also involves knowing how to manage resistance. Effective investigative interviewers ask open-ended questions, stay focused on the answers, pursue contradictions, and listen "between the lines." This type of interviewing is not judgmental toward the interviewee, looking instead for genuine common ground out of a concern to prevent future incidents.

Cause Tree Analysis of the incident. See Chapter 12 for a detailed presentation of the steps of Cause Tree Analysis.

Confirm the Analysis. After developing a Cause Tree account of the incident, the investigator re-scrutinizes the site of the incident to see whether the account is plausible in light of what he/she finds there. This provides a check to see whether the Cause Tree account needs to be changed because of new facts discovered during a visit to the site of the accident. This step sometimes requires making a drawing of the equipment or other facilities involved. It may also require taking samples of equipment or debris for analysis.

Research past incidents. Once the investigator is satisfied that the Cause Tree picture of the the incident is adequate, he/she checks to see whether similar incidents have occurred in the past. The investigator wants to discover what the earlier investigations uncovered, what Prevention Action Plans were developed for the past incidents, did the plans work?

Problem-solving. The unsafe behaviors uncovered by the confirmed Cause Tree Analysis may be further scrutinized using ABC Analysis (Chapter 2) to determine the full range of antecedents and consequences of the behaviors. The investigators brainstorm solutions to each of the chains of causes contributing to the incident. The solutions address each of the chains of causes of the incident—equipment, procedures, new behaviors to be added to the inventory of critical behaviors. The considerations that arise here are the same ones that emerge in the problem-solving stage of the Safety Meeting described above in this chapter. The Action Plan generated by the investigators is very similar to the Action Plan generated by the safety meeting. Below is an Investigation Report, followed by two tables which present, respectively, an analysis of relevant observation data, and an Action Plan drawn up in response to the reported incident.

Supervisor's Investigation Report

Department: Caustic   Date of Occurrence: 10/10/89
Time: 7:08 p.m.   Date Reported: 10/16/89
Employee Involved: George Smith

There was no property damage, but injury did occur, the employee sustaining a chemical burn on his right cheek. Substance was KOH. The employee was talking to a maintenance mechanic, Harry Beese, at the K-12 filter.

Description of how incident happened: George was discussing the line with Harry Beese, who was breaking into a 1-in flange to remove the blind on K-12 filter. George was wearing safety glasses but not goggles. George was not standing to the side but directly in the line of fire. Harry broke the line without asking George to move. Neither man had looked for bleeder, and when site was inspected no bleeder was found between the block valve and the blind. The valve was leaking. It was mere luck that KOH did not contact George's eye.

Behaviors or conditions that contributed most directly to this incident —

- 3 Behaviors: (1) Pre-job inspection, (2) Line of fire, (3) Response to known hazard/communication.
- Conditions: no bleeder, leaking valve.
- Management systems: The facility has no written procedure for installing or removing a blind; at present, no one systematically looks for design problems.

As Table 16-1 shows, behaviors (1) and (2) have been identified and are currently being observed but at a low frequency. The high %Safe for these

Table 16-1. Analysis of behavioral Observation data

IDENTIFIED CRITICAL BEHAVIORS	CURRENTLY ON THE INVENTORY?	%SAFE	%SHEETS
1. Pre-job inspection	yes, Item 3.1	92	2
2. Line of fire	yes, Item 2.2	94	3
3. Response to known hazard/ communication	no		

Table 16-2. Sample Action Plan

ACTIONS	PERSON RESPONSIBLE	DATE COMPLETED
1. Review critical behaviors with both Harry Beese and George Smith.	Investigator	Completed, this date
2. Fix valve. Install bleeder.	Maintenance	Completed, this date
3. Discuss incident with operators and ask them to collect instances of similar design problems for next Safety Meeting.	Supervisor Investigator	In Progress, will be completed within two weeks
4. Write first draft of Blinding Procedure	Harry Beese and George Smith	Due within four weeks
5. Add behavior (3)— *Response to known hazard/ communication*—from analysis to the facility's inventory of critical behaviors.	Steering Committee	To be added within one month
6. Increase frequency of observation for *Pre-job inspection* and *Line of fire*—behaviors (1) and (2) respectively from the analysis.	Supervisor all Observers	To begin immediately
7. Review %Safe for items (1), (2), and (3) in three months. If they are low, do ABC Analysis.	Safety Meeting Facilitator for the (specific date) safety meeting	Yet to be initiated

items is misleading since so little data is being collected on these behaviors. The facility can expect to continue to have injuries like this one until it adequately samples these behaviors routinely. Observation frequency needs to be increased on behaviors (1) and (2). ABC Analysis and Cause Tree Analysis of this incident might be required in order to develop a plan to improve the %Safe behavior once more. Behavior (3) needs to be added to the facility inventory of critical behaviors. These issues are addressed in the Action Plan drawn up in response to this incident (see Table 16-2).

System Issues

In order to make best use of the information, an accident data base is maintained so that it includes incident reports and Action Plans for each accident. A system of follow-up is put in place for Action Plan items, and training is provided for personnel who lead incident investigations.

Chapter 17 offers five brief case histories of this process in action.

Chapter 17

Case Histories

INTRODUCTION

This chapter presents a cross section of case histories. Having worked with Implementation efforts at approximately fifty locations as of this writing, the authors could outline many success stories here, but there is greater benefit for the reader in a description of the range of projects. Therefore, these case histories show not only the efforts that succeeded but the ones that failed and also the ones that stalled and required additional effort in order to succeed eventually. In all, only a few of the Implementation efforts have failed. A number have stalled, and the great majority have been successful.

In a recent development toward the end of the 1980s, various facilities engaged in Implementation efforts have joined the authors in organizing Users Conferences to compare notes on subjects as diverse as performance data and employee involvement. The participants come from a range of industries for the networking and problem-solving opportunities which these semi-annual Users Conferences afford. In effect, the conferences have proven to be an effective means of exchanging information of the type presented here.

CASE HISTORY 1

At the time of Assessment and Implementation efforts, this nylon manufacturing plant in the southeastern United States had 2500 employees. The plant was not unionized. Previous efforts had been made to improve safety through traditional programs, some of which had been successful and some of which had passed almost without a trace. The internal company recordable injury frequency rate ranged between 1.0 and 2.0 and was quite low compared to the industry and to the parent organization as a whole.

The two largest departments in the facility took on the project of Implementing the behavior-based approach to safety. One of these departments was an operations unit with about 1100 employees; the other was a maintenance unit with about 700 employees. During the Assessment effort and before the decision was made to Implement, the consultant provided an introduction of the basic concepts and procedures to a cross section of approximately 200 employees. This introductory training took place over the course of one month. Employees were asked whether the behavioral approach to safety was a method that they could support, and whether they thought it would be wise for the company to spend its resources on an Implementation effort.

This issue was debated and discussed in considerable detail. Early in the Assessment effort an informal Steering Committee was formed. The Committee was made up of key managers and the consultant, and its purpose was to guide the decision-making process and to provide leadership so that employees would be well informed and would feel supported and reassured as they considered the matter before them. Concerns were expressed and recorded, and the general feeling of the group of 200 employees was that it was a good idea to pursue the behavior-based safety process provided that management would commit to seeing it through in the long term.

The Steering Committee also considered organizational issues such as how to be sure that key leaders understood and endorsed the behavioral process, so that they in turn could support it and keep it going. Working closely with the consultant, this informal Steering Committee was very important to the very high level of success of the Assessment effort. It is noteworthy that the informal Committee was not formed intentionally. The Committee came together in the course of meetings between the consultant and key managers who were interested in seeing significant improvement occur

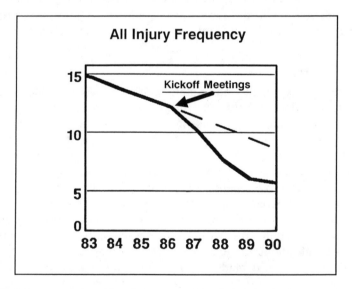

Figure 17-1. Increased rate of improvement resulting from behavior-based safety process at nylon producing plant.

in safety performance and who thought that for this purpose the behavioral approach was a sound one.

Following the decision to pursue Implementation, formal Steering Committees whose members were primarily hourly employees were set up in each of the two departments. The Committee members were selected on the basis of their credibility with their co-workers and their commitment to safety. Supervisors were resource persons for each of these Steering Committees but did not sit on the Committees themselves. The Steering Committees were given adequate time to meet as a group and talk about issues involved in the Implementation. They were trained by the consultant in how to train Observers, supervisors, and managers. The Committees developed the inventories of critical safety-related behaviors for their respective departments, and they conducted the Inventory Review Meetings with their peers in order to get buy-in and endorsement of the behavioral approach.

In the maintenance department the Steering Committee jelled very early and took on a significant commitment to see the Implementation effort through. The Committee members were engaged, motivated, and dedicated. They had the support of their fellow workers, and they became very enthusiastic about making the behavioral process work. This successful Committee had twelve members.

The Steering Committee in the operations department had twenty members. It was too large and not only did this Committee not jell early; it floundered. The larger Committee spent many months working, counter-productively, on the department inventory of critical behaviors. The inventory itself was too large. There was too much unresolved disagreement among the Committee members about the scope of the inventory. In addition, Committee members did not attend meetings regularly. These problems were due in part to the Committee's large size, and in part to its unpreparedness (the members should have had training in group functioning). Training in teamwork, problem-solving, conflict resolution, how to facilitate a meeting, and so forth would have been very helpful if it had been provided to the newly formed Committee.

The successful Steering Committee in the maintenance department had the benefit of a full-time facilitator. This was an hourly employee who took on the task of making sure that everything that needed doing got done. This challenge included inventory development, Observer training, interpretation of observation data, and outreach, publicity and communication. The maintenance department Implementation effort was a substantial success from the beginning. It had very high levels of support and involvement from employees throughout the organization. It achieved a substantial reduction of injury frequency during the first and second year (over 75% combined) and continues to improve at this writing. This Implementation effort was also characterized by a significant change in the safety culture of the department, a safety culture in which a much higher significance was accorded to safety performance. Many "side" benefits also accrued to the department organization—generally improved communications, teamwork, working relationships, and productivity improved simultaneously. See Figure 17-1.

Furthermore, this department provided additional innovation in that the Steering Committee "budded off" to form a number of sub-committees which took on a variety of new tasks. One sub-committee re-evaluated the safety incentive system to improve it by getting away from tangible gimmicks and prizes and moving toward intangible (social) reinforcement for significant contributions to the safety effort. Another sub-committee investigated off-the-job safety. Another focused on department communications and procedures.

Meanwhile the operations department had a small reduction in injury frequency the first year and generally mixed results. It was a struggle to maintain an adequate frequency of observations. It was a continuing struggle to get employee endorsement of the Implementation effort. Getting the Steering Committee to function effectively was a struggle. However, in the face of the difficulties, the operations department management remained committed to the effort. They were supported by the site manager, and they continued to push hard for Implementation. They provided adequate support and appointed two full-time facilitators to make the Implementation effort effective. Gradually, near the end of the first year and moving into the second year, the frequency of observations rose to adequate levels, and the quality of observations began to improve. During the first year, the injury frequency rate declined about 20%; and during the second year, injury frequency was reduced by approximately 50%, with continued improvement as of this writing.

CASE HISTORY 2

At the time of Assessment and Implementation efforts, this chemical company in the south-central United States had a union-represented workforce of approximately 1100 employees. After an introductory presentation to senior management at the site and introductory seminars for key employees, a pilot area was selected for Implementation of the behavioral safety process. The pilot area involved about 110 employees and covered several sub-areas within one department.

A Steering Committee was formed. Its membership included a high proportion of hourly employees. The Assessment showed that complacency was an issue in several work groups—accident frequency was so low that safety was not recognized as an area needing improvement. Plant wide accident frequency was low for the industry but level over several years for the location. Steering Committee members were also designated as Observers. The Steering Committee visited other chemical plants where the behavioral process was in use to determine potential problem areas and to gain assistance with planning.

A second-level supervisor was selected as the Facilitator/Steering Committee chairman and was instructed to use about one-fourth of his time to accomplish this new task. The supervisor was selected for the position because of his unique ability to provide leadership and at the same time motivate hourly employees to take an active and participatory role in the behavioral safety process.

Union support was identified as a key Implementation issue from the beginning. Employee relations had been strained in the preceding several years, and the union leadership was skeptical of management's real commitment to safety over the long term.

The quality improvement process at this facility had been successful on the whole. It was statistically-based and had been sustained and improved over a period of several years. This fact added to management's overall credibility with the workforce. The presence of the quality improvement process also created an opportunity to reinforce quality principles while improving safety performance by emphasizing the similarity in principle between the behavior-based approach and quality improvement.

The authors worked with the facility's SPC consultant to ensure consistency and to take advantage of the opportunity to strengthen the quality effort. Applications of SPC techniques to accident data were explored and developed. Accident data was analyzed within the SPC framework and presented accordingly.

Following the development of a behavioral inventory, the Steering Committee was sent off-site for two days for a team building exercise in which the authors participated. The exercise, which was conducted by in-house personnel and based on previous training, proved to be so effective at improving the Steering Committee's level of functioning that the authors have subsequently recommended it to many other companies.

Several union leaders were selected for additional training to prepare them to be Trainers. This provided an opportunity for them to meet other hourly employees and discuss common issues and concerns. As a result, they were able to support the Implementation effort wholeheartedly.

Frequency of observations at this site were within the target range. %Safe behaviors improved consistently. As shown in Figure 17-2, accident frequency within the pilot areas declined considerably while plant wide accident frequency increased. Note in this figure that accident frequency is shown as number of exposure hours per accident. So achieving more hours per incident represents an improvement in safety. This indicator was selected

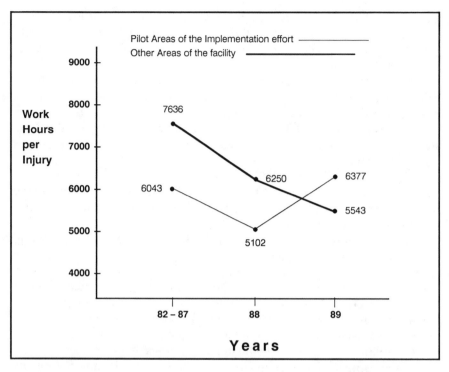

Figure 17-2. Case History 2—within the pilot areas, steadily improving safety performance throughout 1988.

by the company's statistician. This development provided site level management with the motivation needed to Implement the process plant wide. This Implementation effort is ongoing as of this writing.

CASE HISTORY 3

At the time of Assessment and Implementation efforts, this unionized paper mill in the northwestern United States had 700 employees. Labor relations had deteriorated to the point that the union membership was very suspicious of management, and the union leadership was militantly anti-management. An introductory presentation to the mill manager and staff was well received and was followed by a presentation to supervisors and to union officials. Union officials seemed to have their minds made up about the practice of making observations of the workforce—they just knew that it had to be a cover for punitive measures somehow. It was very difficult to discuss this matter with them. They were stubborn, belligerent, and deaf to all reasoned discussion. Their manner, in a word, was adversarial. In spite of this environment, there was enough common interest in preventing injuries to keep the meeting alive.

There followed a meeting with department heads. Although they wanted to do something, everyone disagreed with everyone else about how to proceed. Right in the middle of the meeting, a department head with a workforce of approximately 100 employees simply said, "I want to start this process in my department, and I know I can work with the union." It turned out that he was right. Even in such a poor environment this department head had good relations with most employees due to his unusual communication skills, reputation for honesty, and overall good will.

The Assessment of this one department showed that many hourly employees felt in some ways unrepresented by union leadership, which even the union membership described as "radical." The department had a fairly strong safety culture, with the established idea of "helping your buddy" as acceptable safety-related behavior.

A Steering Committee was formed, consisting of 10 hourly employees and a supervisor. The inventory of critical behaviors was developed by the Steering Committee with assistance from the authors. The Steering Committee itself became the facility's first Observers. The department head acted as Facilitator/Committee Chairman. Observation frequency was excellent from the start, as was the general response of the workforce to being observed. The official position of the union was neutral. They neither supported nor opposed the behavioral safety process.

Employee involvement was high and quite positive. Observation data was used in safety meetings to target areas for improvement and to discuss results. Employees liked the behavior-based process and members of the Steering Committee were quite positive about their work. Accident frequency was drastically reduced. During the first year of the continuous improvement process there was only one recordable injury—compared to thirteen recordable injuries in the preceding year. At the end of that year the department head was promoted and transferred to another mill. Unfortunately, the new head of the department did not understand the behavioral safety process. He was not trained in it, and a year later he let it deteriorate and lapse, illustrating the importance to the process of management's intelligent support and the effect of the lack of management support.

CASE HISTORY 4

At the time of Assessment and Implementation efforts, this unionized chemical plant in the northeastern United States had approximately 800 employees, a history of poor performance in safety, relative to the industry, and an active but stalled quality improvement process. The corporate safety director initiated the Assessment. He wanted a pilot facility and he asked the plant manager whether his site could be used. The corporate office paid the consulting and training fee. This was the first in a series of mistakes. Without the commitment required to sustain the process, the plant manager prematurely said yes to the corporate director's request.

The Assessment findings showed that workers believed that as far as management was concerned, "safety talks, production counts." This view severely affected management's credibility with workers about safety. Middle managers themselves felt that they were currently overwhelmed with conflicting demands—quality, production, cost, fewer supervisors per employee, training, and safety. Basic building blocks for safety were in place—accident investigations, written procedures, safety meetings, accountability—but they were not functioning at a high level. Employee involvement was a stated objective but not a reality except for a few exceptionally motivated hourly employees. A few years before the Assessment, a supervisory observation program had been tried and had left a bad taste in the mouths of both supervisors and employees. It was seen as negative—a "supervisor spy program"—and had not resulted in any net gain in safety performance.

In Case 4 no Steering Committee was used during the initial Implementation nor was a Facilitator appointed. The safety department provided a safety engineer to do some coordination. The quality control department provided another person to help with coordination minimally and the site manager's staff acted as a Steering Committee. The authors provided consultation and training.

The union president received a broad overview presentation during the Assessment, and although he was not enthusiastic about the approach, he was not against it either.

Developing the inventory of critical behaviors was done without the benefit of extensive behavioral analysis of accident reports, since at the time the authors had not yet developed that procedure. Instead the inventory was compiled on the basis of interviews with employees and input from key personnel. Operational definitions were written along with an observation Data Sheet. Adding to the list of missed opportunities, no time was scheduled for Inventory Review Meetings at which to get buy-in for the inventory at the wage employee level.

Training of Observers showed the first strong sign of Implementation problems—Observer-designates attended scheduled training sessions in low numbers, with little enthusiasm and no knowledge of what the process was all about. Response to training was lukewarm.

Kickoff Meetings were conducted by the safety department with a strong statement of support from the site manager. However, the meeting rooms were crowded, visibility was poor, and the room temperature was uncomfortably high. Most of the Kickoff Meeting was given to a review of slides of various critical behaviors. Although this presentation had the potential (and the intent) of communicating vital information to the employees, it was boring to them and met with resistance.

Follow-up or advanced Observer training was never completed at this facility.

Frequency of Observations (the next clear-cut sign of progress) was very low, about 20% of the target frequency. Basically this Implementation effort never became a self-sustaining process but remained a program. As a program it struggled along for about a year, at which time the safety department sent an engineer to a seminar with other locations, to learn more and exchange information. This engineer learned quickly from the other participants that he did not have the support of his plant manager. It is noteworthy that the other seminar participants, who had an outside perspective, could see this easily while the engineer could not. He went back to the plant with a renewed enthusiasm and attacked the problem directly. The result was worthwhile.

A Steering Committee was formed. An hourly employee was appointed to be full-time Facilitator/Chair. The behavioral inventory was rewritten. The union was brought on board, and observation frequency increased. So did the involvement of the employees, and accident frequency started to drop. Additional employees were sent for training, improvements continued to develop, and by the end of the third year a fair amount of progress had been made. Accident frequency declined about 20% in the second year and an additional 45% in the third.

Issues remained at this plant. Observers were not accepted by the workforce in all areas. Middle managers were not uniform in their support of the process. Observation frequency was still uneven. But a foundation had been built which represented real progress toward cultural change and continuous improvements in safety performance.

CASE HISTORY 5

At the time of Assessment and Implementation efforts, this chemical plant in the northeastern United States had 1400 employees, traditional management structure, and a strong union presence.

This facility was actually two plants. In years past they had been owned by different companies, but at the time of this case history they had been consolidated under the ownership of one of the companies. The plant manager of the joint facility had been the manager of one of the plants before. At both locations union relations were strained, and safety was an issue, but this was especially true of one of the plants.

Both the Assessment and the Implementation at this facility were primarily top-down efforts. No Steering Committee was formed, and there was no involvement of hourly employees. The facility manager and his staff decided to go forward with the behavior-based approach, although it was regarded as new and somewhat experimental at that time. The facility inventory of critical behaviors was developed by the consultant. Training was also conducted by the consultant, including the Inventory Review, which was a lengthy process involving the facility's supervisors. Implementation was marginally successful at first but ended up a failure.

The supervisors did endorse the behavioral approach to some degree; however, their skepticism about management's commitment to see the process through interfered with the process itself. On the whole the supervisors were obstructive, nit-picking many details and raising numerous questions at every step. The consultant also trained the facility's Observers, who were second level supervisors and engineers. The Observers did both verbal feedback and charting. However, at that time there was not a systematic way to accumulate behavioral observation data by category—a situation which motivated the authors to develop a computer software program to handle these data base functions.

Extensive Kickoff Meetings were planned and the consultant trained facility personnel to conduct them. However, it turned out that because of lack of time the Kickoff Meeting Facilitators used a 20-minute slide and video tape presentation as the method of introducing the hourly employees to the new behavior-based safety process. The presentation was of poor quality and it was received poorly.

Nonetheless, the facility's internal company recordable rate dropped to 0.7 during the first year of the Implementation effort, down from a pre-Implementation rate of 1.1. At the beginning of the second year, however, an early retirement incentive was announced, and of the second level supervisors who had been trained to do observations and to provide feedback about two-thirds took early retirement. To make matters even more problematic, the facility Safety Manager, who had been the coordinator of the Implementation effort, was transferred and a new person took his place. At some time during these difficulties, the site manager lost heart. A major retraining effort would have been required to keep the Implementation effort going, and the organization had resisted the attempt to get it started in the first place—even though there were good results in terms of accident reduction. Many departmental managers did not support the Implementation effort and felt that they had received an ambiguous message from the facility manager. This was because the facility manager was in conflict about the degree of direction he should give his department managers, and was therefore unable to provide leadership in such a way that they would want to support the behavior-based approach to continuous safety improvement. The result of all of these factors was that on the basis of a committee recommendation the safety process was modified so much that, in effect, it did not continue.

This project failed for a number of reasons.

- Given the amount of resources that managers were willing to commit, it was too ambitious to attempt an Implementation effort for the entire facility of 1400 employees.
- The site manager did not really understand the behavior-based process, nor what would be expected of him in order to see the Implementation effort through to the establishment of the safety mechanism.
- Most importantly, the management approach at the facility was from the top down, and it ran into the problem of all top-down programs— namely, people do not endorse or buy into something that they do not feel involved in.

The problem of employee involvement is a perennial problem in industrial programs. They always have the same result: a program has temporary success and then it fades out, to be replaced by yet another program. This lesson has been learned many, many times but programs continue with the same problems and short lives.

SUMMARY

The remedy to the numerous failings of the program-approach is to stress organizational development. It cannot be too much emphasized that there is no substitute for the important work of *adapting* new ideas rather than simply adopting them. Furthermore, it turns out that this adaptation effort requires employee involvement throughout. This means involving employees in any new initiative, even at the decision-making point. They need to have a voice in whether or not a new approach should be taken at all; and if it is taken, how it should be structured. This is the only way to be sure that the workforce will support a new approach. Although this kind of employee involvement requires more resources and more time, it is a good idea, one that pays off in substantial benefits in a number of areas in addition to improved safety performance.

PART 4. SPECIAL APPLICATIONS

Chapter 18

Self-Observation

SPECIAL APPLICATIONS—AN INTRODUCTION

Chapters 10 through 16 present the Implementation effort as it moves from training, through behavioral inventory development, observation and feedback, and toward its goal of establishing the continuous improvement safety mechanism. As the safety mechanism is established, the work group and its advisors use the procedures and instruments of the continuous improvement process to identify new areas with high potential for enhanced safety performance. This ongoing focus of the work group as a whole is the generic activity of the behavioral safety process as a self-regulating safety mechanism. Practiced by itself, without the benefit of possible special applications of the behavioral safety process, the self-regulating safety mechanism assures continuous improvement of performance. However, there are some important special applications that are available, and for the facility with the commitment to pursue them they offer additional, highly focused results. In this chapter and in Chapters 19 and 20, three of these special applications are presented:

- Self-Observation and Feedback
- Avoiding Multiple Accidents—Assisting the Employee who has had Multiple Accidents
- Behavioral Back-Injury Prevention

Self-Observation. In many industries there are key groups of employees who work primarily by themselves—drivers, for instance. At a facility with a significant number of such employees it can be very helpful to extend the inventory of critical behaviors and the Data Sheet to the special application of Self-Observation. This variation involves the development of an inventory of critical safety-related behaviors on which drivers learn to rate themselves. In this way the facility's drivers measure and chart their own safety performance over time. The body of this chapter presents a sketch of this application of the behavior-based approach to safety improvement.

Avoiding Multiple Accidents. In many organizations a relatively few employees account for a disproportionate share of their facility's accidents and injuries. Therefore it can be very profitable for the organization to focus the continuous improvement safety process on the performance of these workers. See Chapter 19 for more information on this subject.

Behavioral Back-Injury Prevention. Back injuries are expensive. As a source of lost time and productivity, back injuries rate second only to the common cold. Consequently, a facility looking for ways to extend the reach and effectiveness of the behavioral safety process has good incentive to consider a back-injury prevention effort aimed at helping workers assess their own vulnerabilities. This approach involves giving the employees criteria they can use to measure their own progress toward body-use behaviors that protect against back injury and directing the observation process specifically toward behaviors associated with back injuries. Chapter 20 presents the details of such an effort, complete with criteria testing to assess vulnerability to back injury.

Applications to Environment and Industrial Hygiene. In addition to these three established approaches, at the time of this writing, several companies are engaged in applying the behavioral technology described in this book to the wider target of environmental and industrial hygiene issues. The same principles apply and the methods are the same. The inventory of critical behaviors remain pivotal, but its development for these new target areas requires different techniques and considerable additional effort. A significant advantage of this approach is the appeal that it has for management, who see their efforts affecting a broader, and therefore more valuable, target. It seems likely that during the 1990s this approach will come into focus as more companies organize to cover all three areas of safety, environment, and industrial hygiene within the same structure.

SELF-OBSERVATION

In many industries workers are in remote locations, often in very small groups or working alone. Although this situation would seem to rule out the use the behavior-based approach with its emphasis on observation and feedback, such is not the case. In these circumstances what is called for is Self-Observation. This is a process in which employees themselves provide the observation of their own behavior and prepare their own feedback. Self-Observation has been applied effectively in a variety of settings and is appropriate where motivation to improve is present. Since the employee is responsible for the entire process, motivation is essential to success. In situations where motivation is low, training may be used to develop workforce motivation first. People are naturally motivated not to get hurt and, appealing to this basic fact, training shows the employees how behavioral Self-Observation is a tool to help them avoid injury. Even where motivation is low, however, it can be effective for a facility to train-as-you-go and begin immediately with behavioral inventory development. This is because the activity of inventory development generally supplies motivation for safety performance improvement.

The authors developed a Self-Observation process for a metropolitan transit authority where employee relations were extremely poor. Drivers had no traditional "safety motivation" whatsoever because they hated the organization and felt abused by it. Administratively the organization was in chaos—scandals at the top, strife at the bottom, bureaucratic inefficiency throughout. Even in this environment, drivers were easy to motivate *to improve their own safety.*

Analyzing the sources of their injuries in behavioral terms was most effective. Sport analogies were used to provide a framework within which they could view self-improvement as a positive. Since the drivers developed their own inventory of critical behaviors, they developed a sense of ownership about it. This ownership provided additional motivation.

The Behavioral Inventory and Data Sheet

The Self-Observation inventory and Data Sheet developed with the drivers had 34 items organized into 5 categories. The categories were *Equipment, Starting/Stopping, In Motion, Passengers,* and *Operator Body Use.* Each of the 34 items of the inventory had its own operational definition, each definition representing years of distilled experience about how to operate a bus safely in a metropolitan area. The drivers and their supervisors were trained in these definitions.

The *Equipment* category had one inclusive item, whose definition read as follows:

- Equipment. Equipment must be checked thoroughly before leaving the yard. Failure to do so, or using unsafe equipment, will be rated as Unsafe. Brakes, lights, mirrors, signals, windshield wipers, microphone, horn, wheelchair lift, and the signaling device to the driver that the passenger wants to get off—all of these should be in functional order.

The category *Starting/Stopping* had nine items: Scanning, Starting and Stopping, Signaling, Spotting, Mirrors, Jumping the Traffic Signal, Moving prior to Closing the Door, Announcing Stops, and Leaving the Bus Unattended. Here are several of the operational definitions:

- Scanning. The operator must make a visual scan for the big picture before setting the bus in motion.
- Spotting. The operator must stop the bus with the door at the transit authority marker; and position the vehicle parallel to the curb, and close enough to it so that passenger loading and unloading can be

accomplished safely, and so that no cars can pass on the right side of the bus.

- Mirrors. The operator must make a visual check of both mirrors before pulling out after loading or unloading of passengers.
- Jumping the Traffic Signal. Moving the bus into the crosswalk while the light is still red is to be rated as Unsafe.

The category Bus in Motion had 15 items: Following Distance, Visual Lead-Time, Disposing of Problems, Eye Movement, Commitment on Green Light, Covering the Brake Pedal, Speed, Passing and Turning, Signaling, Hands on Steering Wheel, Knowledge of the Route, Amber and Red Traffic Lights, Schedules, Controlling the Steering Wheel, and Railroad Approach and Crossing. The operational definitions of these items included the following:

- Following Distance. The operator must maintain a safe following distance, defined as no less than a two-second distance from the vehicle ahead.
- Visual Lead-Time. The operator must maintain an adequate visual horizon by looking over the top of the vehicle traffic that is immediately in front of him/her.
- Eye Movement. The operator must scan the entire big picture at least once every 10 seconds while moving and at all intersection crossings.
- Covering the Brake Pedal. The brake pedal must be covered by the operator's foot while proceeding through intersections or hazardous situations.
- Knowledge of the Route. Reading a map or instructions, driving too slow for the conditions and/or speed limit, hesitancy, any one of these counts as Unsafe behavior.
- Controlling the Steering Wheel. The operator must control the steering wheel with a firm grip and in a fashion that does not require the use of another object (the cash box, for example) for additional leverage.
- Railroad Approach and Crossing. During the last one hundred feet of the approach to a railroad crossing, the bus must not exceed ten miles per hour. The bus must stop at the grade of the crossing. The bus will proceed in smooth, single-gear acceleration over the crossing when it is clear. This item is to be rated as Unsafe if the operator fails to do any of these actions. (This item does not apply at any exempted crossing.)

The category *Passengers* had six items: Loading and Unloading, Mirrors, Communicating with Passengers, Reporting, and Excessive Conversation.

- Mirrors. The operator must make a visual check of passenger mirrors after loading and unloading.

- Communication with Passengers. The operator must make friendly, visual contact with passengers and speak to each one, briefly answering questions. There must be no abruptness or gruffness in the operator's manner.
- Passenger Position. In an uncrowded bus the operator will not take his foot off of the brake and commence motion until all passengers are behind the yellow line on the floor of the bus. In a crowded bus, the operator will not operate the bus with any passengers in the stairwells.

And finally, the category *Operator Body Use and Activities* had three items: Twisting and Stretching, Distractions, and Horseplay.

- Twisting and Stretching. The operator will position all mirrors to eliminate the need for twisting or stretching the torso or neck.
- Distractions. While operating a bus, it is Unsafe for the operator to pass time by listening to commercial radio broadcasts. The same is true for the use of tobacco or beverages, or holding such things in the hand or mouth.

The Observation Schedule

The observation schedule was that each driver rated his or her own behavior once or twice daily, at midday and/or at the end of the shift. The drivers computed their own %Safe and recorded it on their individual feedback charts. In addition, a supervisor rode with each driver and did an observation once every two weeks. The supervisor's observation was plotted on the driver's individual feedback chart. If the supervisor and the driver gave the driver different %Safe ratings, they discussed it to determine the source of the difference.

A compiled feedback chart of Self-Observations for a group of thirty drivers is shown in Figure 18-1 along with data on the same drivers from supervisor observations. Notice that the drivers' Self-Observation %Safe ratings declined while supervisor %Safe ratings of the drivers increased. After the Self-Observation process had been in place for several months, the drivers were interviewed to determine their perceptions of the approach. They were extremely positive about it. They admitted that they had been skeptical at first. They also admitted that they had been reluctant to give themselves low %Safe ratings at first, even though they knew that they deserved low scores. However, as they learned that the process was not punitive, they gained confidence and began to be more "honest" in their self ratings. They said that even in the beginning when their Self-Observation %Safe scores were inaccurately high, the process made them aware of their

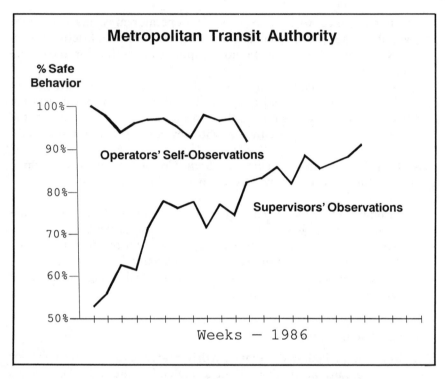

Figure 18-1. Metropolitan Transit Authority.

unsafe behaviors and they were able to work on them to improve. This actual improvement is reflected in the supervisor observations. The drivers reported dramatic improvements in their safe driving abilities, and they were enthusiastic about the behavioral approach.

During the four-month period following Implementation, while the authors were able to track accident frequency it declined 66%.

Chapter 19

Managing Employees Who Have Had Multiple Accidents

Employees who have more than one accident in a two- or three-year period account for a disproportionately high share of total accident frequency in most plants that the authors have studied. It is not unusual to find that 6%-10% of the workforce accounts for 45%-65% of accident frequency. Assuring the safety of these individuals is a challenge. This chapter presents a behavior-based method to help these individuals avoid future injuries. In many organizations helping an employee who has had multiple injuries is usually the job of the supervisor; however, at some facilities this may be the responsibility of the safety representative, or of a designated wage-level employee.

SELECTION CRITERIA

Selecting employees for this special application of the behavior-based safety process is a complex issue. If the selection criteria include first aid injuries, the result may simply be to select those employees who follow regulations and report all injuries. Selecting by more serious injuries—recordable injuries, for instance—eliminates most of this problem. However, since drafting selection criteria for a facility is really a problem in statistics, it is best to get help from a statistician rather than to trust to hunch or common sense in setting the criteria. Among other factors, the statistician will take into account the overall accident frequency at the site. Generally speaking, two or more recordable injuries over a three-year period is a good selection criterion for a facility with an accident frequency rate below 3.0.

Employee Profile

A person with multiple injuries is not necessarily a problem employee. It used to be thought that there was such a thing as an "accident-prone personality"—a person who unconsciously wanted to get hurt. Individuals who had repeated accidents were thought to have such personalities. However, more recent research indicates that this is not the case. It turns out that everyone is susceptible to accidents. This susceptibility probably occurs during times of high stress or great change. Although the precise mechanism has not yet been identified, it is likely that at such times it is not so much that people are prone to accidents as that they may become prone to acting impulsively. In other words, times of high stress and great change are not personality traits but conditions that everyone is subject to from time to time. Other factors that may bear on repeat accidents are mismatches between job and worker, and individual physical vulnerabilities.

The moral here is that people who have had multiple accidents are not bad employees. In fact, judged in terms of conscientiousness and a general willingness to work hard, they are often among a facility's best employees. In any event, they are employees who can benefit from an approach to their safety performance that is positive rather than judgmental and that aims at discovering the causes of their individual accidents so that they can prevent them in the future.

Strategy

In most facilities it is the responsibility of the supervisor to work with the employee on accident prevention; however, as it was pointed out in the introduction to this part of the book, this task may be fulfilled by a safety representative or someone else who is experienced in the continuous improvement process. For the sake of simplicity, the rest of this chapter refers to the *supervisor*, but the material presented here applies to anyone who is supervising the accident prevention process with the identified employee.

The strategy for assisting the person who has had repeat accidents is for the supervisor to form an alliance with him or her, and then together to formulate an Action Plan and a behavioral observation Agreement. The alliance is important. The supervisor and the employee must be allied in order to discover the causes of the individual's accidents and to develop a plan to prevent future accidents. The underlying premise of this approach is that no one wants to have accidents and that together the supervisor and the employee can identify the employee's critical unsafe behaviors which, if altered, will improve his or her chances of avoiding future injury. This approach is non-punitive, therefore. At its core it requires both cooperative exploration for causes and mutual problem-solving. The product of this approach is a behavioral observation Agreement to prevent future accidents.

Steps

The five steps of the procedure for avoiding multiple accidents are the following:

1. The supervisor sets the stage by scheduling an appointment with the employee.
2. The two meet and investigate the employee's accidents.

3. Together the supervisor and employee design an individualized inventory of critical behaviors and a Data Sheet.
4. They formulate a behavior observation Agreement which specifies what each will do to help the employee prevent future injuries.
5. The supervisor and employee carry out their Agreement and meet periodically to see how the Agreement is working.

Setting the Stage

Effective supervisors take the initiative. They approach the employees to set up a time and place for an appointment, explaining that since the employee has had multiple accidents in the last several years, there is a likelihood of re-injury, and that the meeting is to work on a plan to avoid future injuries. It is very important that the employee understand the purpose of the meeting—it is not a counseling session, and it is not punishment. The supervisor might well say something such as the following:

- *When we analyzed our accidents we found that certain people have gotten hurt a lot. They are more vulnerable to injury, so we are starting a new process to work with these people to help them prevent future injuries.*
- *This is not punitive or negative, and it will not go on your record. What we will be doing is looking at your past accidents in order to really figure out what might help. I want to ask you some questions about your accidents to get us started thinking about this together.*

The appointment itself should follow soon after this initial conversation.

For the appointment, the supervisor chooses a place free from distractions, setting aside the two hours of time that may be required during the meeting. This is time well spent if it prevents another injury. In some cases one meeting is not enough. Since this work requires attention and mental concentration, subsequent meetings should last no more than two hours each.

Skills Used in This Process

Certain skills are required in order to make this process work. They are Managing Resistance to Change, Investigative Interviewing, ABC and Cause Tree Analyses, and Pattern Search. Since each of these skills has been discussed in preceding chapters, they are only reviewed briefly here.

Managing resistance.   Quite frequently people resist answering a question they have been asked. This may be due to a bad attitude but more often it is because they either do not understand the question or the reason behind it. Sometimes they have simply wandered from the point, or they have things to say that are much more interesting to them than the question posed by the supervisor who is investigating their accidents with them. These different ways of not answering a question have a common purpose—to side-track the conversation. They represent instances of resistance to change. For a detailed discussion of the various forms of resistance to change and how to manage them, see Chapter 11. Briefly stated, the rule for managing resistance is to expect it and to be undistracted by it.

Investigative interviewing.   Investigative interviewing is also treated at length in Chapter 11. The important variation for one-to-one circumstances such as this is that effective dialogue is very important.

Dialogue. Dialogue is a conversation in which both people learn something. In this case the supervisor and the employee need to learn some very important things about the employee's observable work patterns. They are engaged in an effort to discover the unsafe behaviors which predispose the employee to future injuries. Dialogue comes naturally to them, when both of them are mindful of the stakes.

In addition to maintaining the conditions for dialogue, the supervisor:

- Asks open-ended questions
- Listens for promising leads
- Explores contradictions in the employee's answers
- Elicits all pertinent details
- Notices when the employee changes the subject
- Closes by checking with the employee to see that his/her understanding of the employee's answers is accurate

Each of these topics is covered in Chapter 11. Perhaps the most important subject to recapitulate briefly here is the importance of *Listening.* Supervisors who are working with employees to help them avoid future injuries need to have good listening skills—skills not always associated with the supervisory role. To the inexperienced supervisor, listening may seem contradictory to supervising. Actually, understood properly, good listening is a prerequisite to good supervising. Effective listening does not require *agreeing with* what another person is saying. What it does require is an effort to *understand* what another person is saying. Supervisors with very strong opinions of their own to start with may have a hard time doing this—their own opinions or judgments get in the way of their ability to simply listen so that they understand what the employee is saying. Understanding the employee's descriptions of his or her own accidents, however, is a prerequisite to developing an individual Action Plan that helps to change unsafe work patterns. Briefly stated, effective listening is:

- Non-judgmental
- Addressed to common ground
- "Slow witted"
- Persistent
- Friendly
- Responsiveness
- In charge

Pattern Search. ABC Analysis, Cause Tree Analysis, and Pattern Search are helpful techniques for investigating an employee's accidents. For a presentation of ABC Analysis, see Chapter 2, and for Cause Tree Analysis, Chapter 12. After completing Cause Tree Analysis of the employee's accidents, the next way to focus the conversation is to search for patterns of behavior that are common to the accidents. Pattern Search can be quite subtle, consider the following three pairs of accidents for what each pair has in common.

Worker A. Chemical splashed in his face when he jerked loose a sample-container that was stuck; and he smashed his knuckles when he was loosening a nut on an old housing. *Common Pattern:* something that was stuck suddenly gave way and the worker's body was in the *Line-of-fire.*

Worker B. Dust got in his face and eyes when he was changing an overhead light bulb; and foreign body got in his eyes when he was checking a grind. *Common Pattern:* something got into the worker's face and eyes. Was he too close? Does he need corrective lenses? Are his eyes exceptionally dry.

Worker C. He fell when his ladder wobbled on slightly uneven ground; and he twisted his ankle when he stepped off a platform onto clutter. *Common Pattern:* something was assumed to be stable and then gave way. Did he fail to foresee potential hazard? Is his sense of balance less acute than it needs to be when the worker deals with situations involving unstable footing? Are the worker's muscles so tight that even small lurches are liable to pull them?

The supervisor is also sensitive to other factors such as common circumstances, common body part injured, and common type of activity.

Common circumstances. The supervisor looks to see whether the worker's accidents involve situations that are similar. Such situations might include working in a confined space, or in a cluttered area, in poor lighting, or at heights.

Common body part. Does the worker have a pattern so far of injuring his face, back, hand, etc. Sometimes he may have injured different body parts but the same tissue—a pulled shoulder, a pulled back. Both of these injuries involve ligaments and tendons.

Common activity. Perhaps the worker was injured each time while doing the same thing—climbing, carrying something, opening something, etc.

Developing the individual inventory. Cause Tree Analysis and Pattern Search are methods that the supervisor can use to focus the meeting with the employee. By using these behavioral techniques of analysis together they discover the multiple causes of the worker's injuries. The causes may well involve a range of factors in addition to individual behaviors and vulnerabilities, factors including facility and equipment, peer culture, and management systems. Having discovered the causes of the employee's past accidents, the next step is to develop an inventory of behaviors with the employee to help him or her manage the identified work patterns.

The subject of the inventory of critical behaviors is treated in detail in Chapter 12. Here, just as in the inventory for a facility or a department, the inventory for an individual worker is a list of items which can be used to take a measure of safety performance. In this case the measure concerns the safety performance of just one person. The individual inventory for the employee and the wider inventory for the facility also have the following points in common:

- Worker input into the inventory is crucial to its success
- The inventory items;

    - do not overlap,
    - are of Observable things only, and
    - are operationally defined.

For an explanation of any of these key factors, see Chapter 12.

Behavioral Observation Agreement

At this stage of working with the employee, the supervisor is very close to doing actual observations using the inventory of critical behaviors that the supervisor and the employee have developed together. It is at this point that the two need to draw up a behavioral observation Agreement. This Agreement has some very important purposes, and as a method of clarifying the safety improvement process it has proven advantages. The observation Agreement is in effect an Action Plan, and like a business agreement it has two sides to it, one setting out what the supervisor agrees to do, the other setting out what the employee agrees to do. Both sides need to be spelled out in sufficient detail so there is no confusion about them.

The underlying strategy is for the supervisor to agree to remedy the management and equipment issues (if they have not already been attended to) and to provide special feedback to the employee for the behavioral issues that they discovered when they analyzed the employee's accidents. In other words, the observation Agreement takes into account all of the factors discovered during their mutual investigation of the employee's accidents, including facility conditions and the employee's special vulnerabilities. The individual inventory also addresses these factors. Taking these things into consideration, the Agreement assigns responsibility for observations and a deadline for any unresolved management and equipment issues.

In committing oneself to an Agreement, the employee might agree to the following: I will do an overall inventory rating on myself at least once each day, and it will include my personal inventory items which are:

- Wear the corrective lenses that have been prescribed for me
- Position my face out of the line-of-fire
- Keep face protection with me at all times, and use it when it is required.

As the other part of this same Agreement, the supervisor might agree to the following:

- I will do an inventory observation on the employee twice a week, including her individual inventory. I will chart my observations of her and give her verbal feedback on her safety performance.
- I will have all members of the crew inspect tools and equipment for one week, and during that time I will write up and replace any tools that are defective.
- I will also see to it that the bent sample-container receptacle is repaired within one week of this Agreement date, and I will follow-up promptly on equipment problems identified by the crew.

Agreement follow-up. Like any Action Plan, this one requires follow-up. both the supervisor and the employee must keep their respective parts of the bargain. The supervisor continues the observation and feedback schedule called for in the Agreement until it is clear that the worker has developed a consistently high level of safe behavior. The supervisor then cuts back the frequency of observation and feedback and follows a less predictable schedule. Behavioral observation and feedback are central to the Agreement. Observation is an important subject, treated at length in Chapter 14. The point here is that the effective Agreement establishes an *expectation* on the part of the employee that he or she will *receive* individualized feedback on safety performance. By the same token, the Agreement establishes an *obligation* on the part of the supervisor to make observations and to give feedback. There may still be occasions when the employee does not want to hear any feedback about performance. There may be days when the supervisor does not particularly want to make observations and give feedback. Nonetheless, the Agreement makes it clear that this observation and feedback is an inherent part of the process.

From the supervisor's point of view, this one-on-one observation and feedback with the employee is an instance of Behavioral Coaching for Skills Development, a topic treated at length in Chapter 11. The supervisor follows the coaching cycle outlined there. Stated briefly, this means that the supervisor is careful to verbally recognize and respond to even a slight improvement in the employee's performance. This is because some behaviors are very hard to change. Therefore the more sensitive the supervisor is to any movement in the right direction, the better it is for the employee. This sensitivity also affects the order of the supervisor's remarks, responding first to improvements, however slight, and only then making suggestions for further improvement. The supervisor does not criticize poor performance, but provides positive suggestions for improvement.

SUMMARY

As is probably apparent from the various cross-references in this chapter to other chapters of this book, this special application of the behavior-based process covers the entire Implementation effort but in a one-to-one situation.

Chapter 20

Behavioral Back-Injury Prevention

INTRODUCTION

The special application presented in this chapter is the use of behavioral techniques for back-injury prevention. The purpose of behavioral back-injury prevention is:

- To focus the behavioral Observation and feedback system on back injuries in particular, and
- To educate and motivate participants for the correct behaviors to ensure good back care.

Training in the approach presented here enables supervisors and others with specialized areas of responsibility to help employees avoid back injury or re-injury. Wage employees in turn need to know how the back works as a system, and which behaviors are critical to back care. They need to know what is required for proper back use and how to test themselves to see whether they meet these minimum requirements.

This chapter teaches the parts of the back and how they work. It covers such subjects as what gets injured in a back injury, who gets injured, how to prevent injury, and a self-test that employees can use to see whether they have the critical behaviors that make up back fitness.

The Motivation for Preventing Back Injuries

Back injuries are expensive, and they cause significant human suffering. It has been estimated that in the United States back injuries and their associated expenses cost from $12-$14 billion annually. According to one source, one cent out of every 25¢ worth of postage goes to treat the back problems of U.S. Postal Service employees. Furthermore, an employee who is off work with a back injury puts a strain on the rest of the organization which must adjust for the absent worker.

From the employee's standpoint, of course, back injuries are painful and debilitating. Studies indicate that back injuries are so stress-provoking, in fact, that a person who suffers a continuous back injury for six months or more has a 70%-90% chance of becoming clinically depressed, requiring treatment for depression as well as pain. Back injuries can be disabling. They can mean major changes in the way the sufferers live, severely limiting their activities either from chronic pain or from fear of re-injury. Diagnostic procedures for back injuries are themselves painful and can have serious complications. And treatment for back injury may result in addiction to pain medication, muscle relaxants, and sleeping pills.

On the positive side, learning the right behaviors, getting in shape, and using body mechanics in the right way can be a pleasant and reassuring challenge for workers. It can even give them more energy to do other things that they enjoy at the end of the work day. It also means that people can use their work as an opportunity to stay fit.

Who gets injured?   For people who are 20 years old or younger, the odds of suffering back pain are about 50:50. People who are 30 years old or older stand a 90% chance of suffering back pain. At some point in their lives 60%–80% of all adults have a significant complaint of back pain. Back trouble is the major cause of disability in people under 45 years of age. It is second only to the common cold in terms of lost time from work. For people who have already had a back injury, the odds increase three to ten times that they will be re-injured. In other words, the odds are high that most people will experience back pain at some time in their working life. The key to prevention is behavior.

Pain versus injury.   Back pain and back injury are not the same thing. Pain merely signals the need to attend to something. Pain may be due to injury, but not necessarily. Therefore pain is not of itself evidence of damage to the back. Injury on the other hand, means that tissue has been damaged. Many people confuse these two things, thinking that they have a "bad back" when all that they really have is back pain. The most common cause of back pain is muscle fatigue. Muscle fatigue is simply a sign that there has been a build-up of lactic acid in the muscles because they have been more active than they are used to being. This is a normal process, and it is a sign that the muscles are beginning to get into shape.

The most common actual back injuries are strains, sprains, herniated disks, and osteoarthritis.

Back injuries can be prevented.   With age everyone progresses toward either arthritis of the spine or herniated disks, and it might be true that if all people lived long enough, they would develop back problems. However, this degeneration with age is tremendously accelerated or slowed by how people behave and how they use their bodies. In fact, not all people develop back problems as they get older, which is proof that back problems can be prevented.

Informal studies of older workers who had never had back problems despite the fact that they worked in jobs requiring strenuous physical labor showed some important things. These fortunate workers had three behaviors and results of behaviors in common with each other:

1. They were in good physical condition.
2. They were properly flexible.
3. They used proper body mechanics.

The authors discuss the conditions for each of these three factors below, but first a related question. Why is it that even though most people have been taught to lift properly, they do not always do what they have been taught? It turns out that there are naturally occurring consequences that favor unsafe behavior, and these same consequences determine people's behavior unless they know about the consequences and do something to counter them.

Bones, disks, and ligaments sustain the injuries. It used to be thought that the muscles were the parts of the body that were injured in back strain. Nowadays this appears not to be the case. Apparently it is the bones, disks, and ligaments of the back that are usually damaged when the back sustains injury. The pain was in the muscles, and this made people think that the injury was in the muscles, but it has been discovered that the muscle pain associated with back disorders is usually due to fatigue, not to damage. This is a very important point. Back injury is known to be such a debilitating thing that pain in the back can cause people a great deal of fear that they have suffered a serious injury. As a consequence of their fear, they may become ensnared in the litigation-compensation disability syndrome. This can be very damaging to the worker and to the company. However, there is no need to be anxious about back pain if it is due to muscle fatigue alone. Fatigue is a sign of not being in good physical condition. What is needed is not a doctor but proper exercise.

A real back injury, on the other hand, does require the attention of a doctor. The doctor would diagnose it and try conservative treatment which might include anti-inflammatories such as aspirin and rest. But even so, as the patient recovers, the doctor would also prescribe exercises in order to get the back into better physical condition: exercise is an important part of treatment.

The parts of the back. It is useful to think of the back as having two different kinds of parts—active and passive. Passive parts are the ones that require no energy in order to do their job. They do not give one the sensation of effort and they do not undergo fatigue. It is the passive parts of the back that sustain injury through wear and tear. Repetitive motion, high impact activities, and moving things that are too heavy are all behaviors that contribute to this wear and tear. Conversely, minimizing these behaviors minimizes the wear and tear on the back. The active parts, on the other hand, are the ones that do use energy and exhibit stress in feelings of fatigue. These are the parts that do the moving. They get tired, but they are rarely injured. The active parts are strengthened by exercise.

Passive parts of the back. The passive parts of the back are the bones, disks, and ligaments. These are the parts of the back that are damaged in back injury.

The backbone consists of 24 separate bones or vertebrae. Their function is to support the body and to protect the spinal cord and nerves. The numerous parts of the backbone allow it to turn and bend. It is these 24 vertebrae that develop osteoarthritis with age.

The disks are the soft, gelatinous cushions that reside between the 24 bones of the spinal column. They act as transducers, converting vertical shock waves into horizontal forces that are then dissipated into the soft tissues of the body. The disks dry out with age, making them less flexible and functional. This also makes them more prone to injury. The disk fluid itself is highly irritating to the surrounding body parts. When the disks herniate, this fluid is squeezed out into the surrounding tissue, causing an inflammatory reaction. The chance of suffering herniated disks can be minimized by avoiding behaviors that increase the pressure on the disks. This is especially important for older workers, whose disks are more brittle.

Ligaments and tendons are connective tissue. Ligaments are the gristly parts that bind the bones together. Where these parts are concerned, repetitive motion may cause inflammation, carpal tunnel syndrome, or tendonitis, for example. They also can be pulled and torn, resulting in a sprain. Anything that tightens the ligaments and tendons predisposes the back to sprains. These factors include age, cold weather, and important behaviors such as exercise and taking stimulants, even coffee. People who arrive at their retirement years without back injury and with a low chance of back injury, tend to be employees who are properly flexible. This means that they had "loose" ligaments and tendons. See below for a discussion of how to test behaviorally to determine whether someone is properly flexible.

The active parts of the back. The active parts of the back are the ones that burn fuel to do their work. They are the muscles and nerves. The muscles of the body work in pairs; when one of the paired muscles contracts, the other relaxes. The muscles of the back are long straps that run down the sides of the backbone. They are very strong and active. The activity of these muscles is to contract or shorten in length. Their lengthening is a passive process which occurs whenever the muscles they are paired with are contracted. The advantage of having paired muscles is that it makes us highly coordinated—each muscle balancing another muscle and allowing for very fine movements. For example, when the triceps muscle contracts it straightens the arm at the elbow. The triceps is opposed by the biceps which, when it contracts, bends the arm at the elbow.

The opposing muscles of the back are the abdominal muscles. In order for the abdominal muscles to function properly in opposing the back muscles, the abdominals must be very strong. Unfortunately, in most people the abdominal muscles are not as strong as they need to be. This is why doctors often prescribe abdominal exercises for people who have suffered a back injury. In effect these exercises are behaviors aimed at strengthening the stomach muscles, and the reason for them is that it takes a strong belly to work with the back and protect it. Workers who make it to retirement without back injury have a good strong set of back muscles *and* abdominal muscles, as well as certain other muscles. See below for a discussion of how employees can test themselves for the strength of the muscles which are critical to back-injury prevention.

Spinal architecture. There is another element of the back's anatomy that is very important. This is the spinal architecture. The backbone is not straight: there are three natural bends or curves in the backbone. These forward and backward bends in the spine assure that the nerves are maximally protected and that the wear and tear on the disks is minimal. Loss of the natural curves of the spine produces excessive pressure on the edges of the disks, wearing them out faster. Thus, having good body mechanics can be summarized as having behaviors that maintain the natural curve of the spine. These behaviors are:

- Maintaining proper body weight
- Maintaining sufficiently strong abdominal muscles
- Having sufficiently flexible hip joints
- Using proper body movement

Principle behaviors of proper body use. The principle behavior of proper body use is the maintenance of the spine's natural curves. In terms of posture this means that the extremes of a "military bearing" and of slouching are both out. The proper way to sit or stand is with the one foot up on something. This unlocks the hip joint, letting the back adjust itself naturally to maintain its curves. Furthermore, whether sitting or standing it is best to lean so that the spine does not support the body's weight all by itself. Sitting at a desk, for example, the body's weight is supported by the elbow either on the arm of the chair or on the desk top itself. Ideally, when sitting at a desk, the angle between the back and the thigh is about 105 degrees. The chair is for lower back support, and once again it is helpful to put one foot up on something.

Bending behavior. It is important to minimize bending. When a person bends all of the forces are focused on the disks in the lower back. In adults, 60%–70% of body weight is above the waist. Consequently, when people bend at the waist, the passive parts of the backs—ligaments, bones, and disks—have to support whatever they are lifting plus a large part of their own body weight. The physics of bending activity is as follows. The moment of force exerted at the low disks of the back equals the weight lifted multiplied times the distance along the backbone that the weight is from those disks. For instance, when a 200 lb man bends over to lift a 50 lb box, the force at his lower back is actually 4,560 inch-pounds. That figure equals 24 in. (the distance from his hips to his shoulders) multiplied by the sum of the 50 lb box plus 140 lb (70% of the man's own 200 lb body weight). Needless to say, 4,560 inch-pounds is a great deal of force for the man's disks to have to sustain. In fact, this force is enough to damage most people's disks. The formula indicates how important it is to hold the load close to the body as possible.

In order to hug the load close to the body and eliminate bending at the waist, the worker has to bend at the hips. This requires squatting, which uses the leg muscles instead of the lower back muscles. Squatting is an "unnatural" way to lift in that it requires considerable strength and flexibility, and it also requires the activity of active parts. Nutrients are burned to provide the energy for this kind of lifting. Real work is being done to lift in this way, and if workers do not have strength in their legs and abdomen or are not in good cardiovascular condition, they are not able to sustain this kind of lifting effort for very long. It is this fact about the body that brings up the natural consequences that influence people *away* from proper lifting behavior. As was mentioned above, most people know the right way to lift, but do not always act on what they know. The reason being that it requires more energy while the lifting lasts and it takes more work. If a person is not flexible, this way of lifting feels "unnatural." It tires a person out faster, and unless one has sufficiently strong legs it is too hard to do for any length of time. In fact, leg strength is so important for back health that the strength of the leg muscles is a predictor of who will suffer back injuries. Therefore, the authors developed methods presented in this chapter to test a person's leg muscles for adequate strength.

Twisting behavior. The worker also needs to eliminate twisting behavior to avoid back injury. Since the spine is anchored at the hips, twisting from the shoulders means that the torque on the spine adds up all the way down the backbone and is focused on the lowest disks. Above all, it is necessary to avoid twisting and bending from the waist at the same time. This is the motion that causes most back injuries. The way to eliminate twisting is to turn instead. Turning means leading with the foot in the direction of the movement and not with the shoulder. When leading with the foot, the hips follow, and then the shoulders. In this way there is no torque applied to the backbone. Turning requires more energy than twisting, however.

Use the abdomen. The worker needs to use the belly by locking the abdominal muscles. The effect of this is to increase the pressure on the inside of the abdomen, and this pressure in turn pushes back on the bulging disks, lending them support. By tightening the abdominal musculature the exertion of the back muscles is opposed and better coordinated. And finally, if the abdominal muscles are locked, the body will not twist. The torso turns as a unit. Locking the abdomen does not mean pushing the belly out and bearing down. On the contrary, it means tightening the abdominal muscles which flatten the belly making it bulge out on the sides. Locking the abdominal muscles also does not mean holding the breath. It is necessary to breathe normally since holding one's breath decreases the available oxygen, and this causes the muscles to fatigue more rapidly.

Push, do not pull. One reason to push instead of pulling is that while pulling something, a person twists around in order to see better. Further, workers have less mechanical advantage in most pulling positions and are therefore more prone to over-exert themselves. Finally, once the load gets going workers are often tempted to slow it or stop it with their bodies—not a safe practice.

Where pushing is concerned, there is a safe and an unsafe way also. The safety rule is that if a load is too heavy to push straight on without putting the shoulder to it, then it is too heavy to push. In other words, even though a more efficient push develops by putting the shoulder to a load, it is not good for the human back to do so.

Be careful of impact. No sudden starts or stops. The worker's movement needs to be smooth and graceful. Any movement started has to be stopped. The body must take the impact of the deceleration. Consequently, the person who works safely allows plenty of time to stop a motion. It is easy to stop something going 60 miles per hour as long as it is stopped over a long distance. Just as with our car it is unacceptable to stop it by driving it into a brick wall, so with the human back there needs to be a gradual slowing and change of direction. This means that jerking and yanking movements are avoided. The worker who pays attention to this principle will improve coordination—smooth and easy, no jerky stops and starts.

Range of work. Range of work refers to the worker's optimum range. Work materials and tools must not be placed neither too high nor to low. The work should be close to the body and be held between the level of the navel and the mid-chest.

Rest. The worker needs to minimize wear and tear by decreasing repetitive motions. It is important to use different passive tissues to do the same job and occasionally to do the opposite kind of motion altogether. The worker who has been stooping needs to stand up and stretch and vice versa. The principle is to provide periodic relief for the passive tissues that are involved in the repetitive motion. This includes leaning. The worker spares his body and makes other things take the wear and tear.

Warm up. Body parts are more easily injured when they start cold. The blood needs to be circulating, the muscle tissue needs to be warm. This is particularly necessary in cold weather which tends to tighten the muscles of the body and, thereby tightens the tendons too. Also, caffeine intake means that the body requires more warm-up time.

BACK-INJURY CRITERIA TESTING

In order to help people to determine whether they have adequate strength and flexibility of critical muscle groups, the authors have developed a set of criteria tests. Figure 20-1 shows the Criteria Testing score sheet. The first time that the individual does the exercises, the score sheet is used to record the baseline performance. On subsequent occasions progress on the strength and flexibility criteria tests are scored and charted. On all of these criteria-testing positions and exercises, it is important to take the correct position and to avoid strain.

The Human Chair

See Figure 20-2.

Instructions. Have someone brace your feet so that they do not slip out from under you. Sit with the thighs horizontal and lower legs vertical so that the knees are over the balls of the feet. The feet should be from 6-12 inches apart.

Criterion. Meeting the criterion for this exercise requires holding the Human Chair for three minutes. The person who can maintain the Human Chair for three minutes without undue strain has sufficient strength (in the quadriceps muscles of the legs) to correctly use his legs to protect his back.

Warning. If a person has had a previous knee injury or a problem with the knees, permission should be obtained from a physician before doing this exercise.

Corrective exercise. If the criterion cannot be met, exercise the upper legs by assuming the Human Chair position and holding it a little longer each day until it can be held for three minutes.

Employee Name (print) : _____

Criteria – Testing   Scoresheet

Exercise	Criterion		Results			
		Dates :				
Human Chair	3 Minutes					
Sit – Up	3 Minutes					
Hamstring : R	80 degrees	Y / N				
L	80 degrees	Y / N				
Abductor : R	Depressable	Y / N				
L	Depressable	Y / N				
		Dates :				
Human Chair	3 Minutes					
Sit – Up	3 Minutes					
Hamstring : R	80 degrees	Y / N				
L	80 degrees	Y / N				
Abductor : R	Depressable	Y / N				
L	Depressable	Y / N				
		Dates :				
Human Chair	3 Minutes					
Sit – Up	3 Minutes					
Hamstring : R	80 degrees	Y / N				
L	80 degrees	Y / N				
Abductor : R	Depressable	Y / N				
L	Depressable	Y / N				

Your Proper Exercise Pulse Range :
Your Upper Limit = 190 minus your age.
Your Lower Limit = your Upper Limit minus 25.

Figure 20-1. Criteria-testing Scoresheet.

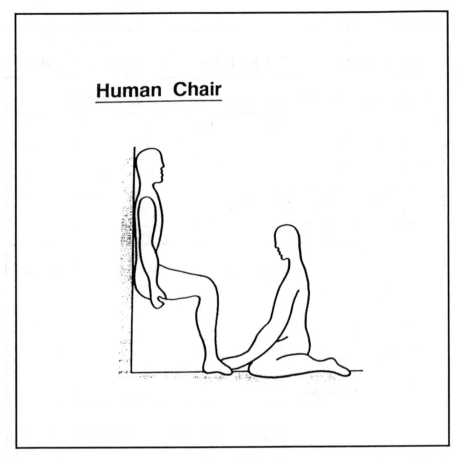

Human Chair

Figure 20-2.

The Partial Sit-Up

See Figure 20-3, Abdominal strength, positions 1 and 2.

Instructions. Start from a sitting position, position 1 in the figure, and have a partner hold the feet on the floor. The knees should be bent, and the body should drop back half-way so that the back is 45 degrees off the floor, position 2 in the figure.

Criterion. Meeting the criteria for the abdominal muscles requires maintaining a partial sit-up for three minutes. It is important to take the correct position and to hold it without strain. If this position can be held for three minutes it means that the test-taker has sufficient abdominal strength to use the abdominal muscles to protect the back.

Abdominal Strength – 1

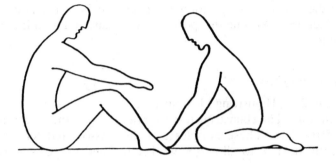

Abdominal Strength – 2

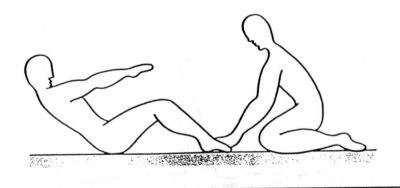

Figure 20-3.

Warning. Start from a sitting position, not from a lying position. Do not jerk or twist. When through, use the arms to lower the body all the way to the floor and get up by rolling onto one side and then standing up. Obtain permission from a physician before doing this exercise.

Corrective exercise. If the position cannot be held for three minutes, hold it for as long as possible, and then let the body down to the floor gently. Record the time that the position was held, and use that as the baseline for future exercises. Assume this position every day, and hold it a little longer each time.

The Hamstring Test

See Figure 20-4, Hamstring Flexibility.

Instructions. The Hamstring test is for flexibility of the long muscles of the back of the legs. These muscles and their associated tendons must be sufficiently flexible to allow the worker to bend properly from the hips.

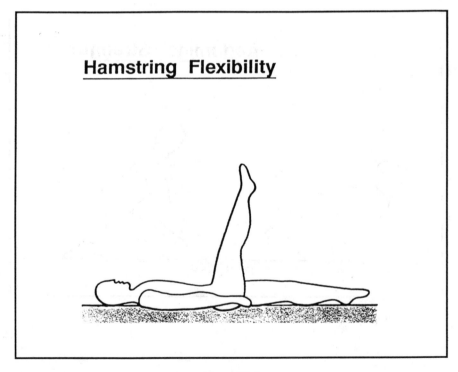

Figure 20-4.

While lying flat on the floor, press the small of the back against the floor. Keep both knees locked, and first raise one leg and then the other.

Criterion. The test here is to be able to raise each leg off the floor 80 degrees while keeping both knees locked.

Warning. The raised leg should not be jerked. Raise it gently.

Corrective exercise. The hamstring stretch, see Figure 20-5.

The Hamstring Stretch

See Figure 20-5.

Instructions. Stand facing a chair or low desk. Raise one leg, and put it on the object. Keep both knees straight. Gently lean forward, bending at the

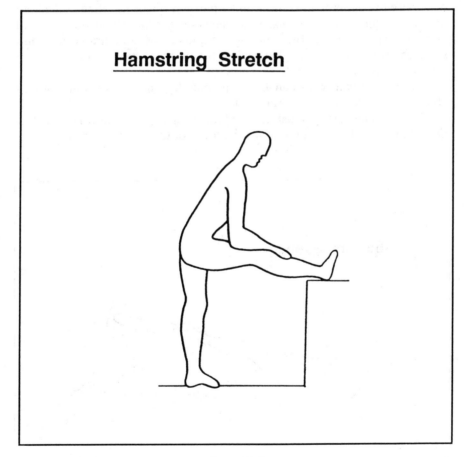

Figure 20-5.

waist and resting the weight of the torso on the raised leg through the arms. Do not hold the weight of the torso with the back. Bending forward a stretch is felt in the back of the raised leg. *Do not bounce.* Instead, gently stretch the back of the leg to the point that it hurts slightly and then hold that position while consciously trying to relax the leg. Repeat this frequently, gradually stretching the hamstring of whichever leg fails the Hamstring Test.

Warning. Do not bounce.

The Leg Abductor Test

The final test is for the leg abductor muscles. See Figure 20-6, Abductor Strength.

Instructions. Lay on one side with the pelvis vertical and the legs outstretched. Now raise the upper leg. At this point a partner, who is kneeling at the feet, attempts to push down the raised leg using only the force of an outstretched arm, not body weight. Both people should be of comparable size and strength. Try to hold up the leg against the partner's push. Both knees are straight at all times.

Criterion. People who can hold up their legs have sufficiently strong abductor muscles to protect the back.

Warning. The partner must not use body weight to push down on the leg. One should not be twisted to resist the force of the partner's arm.

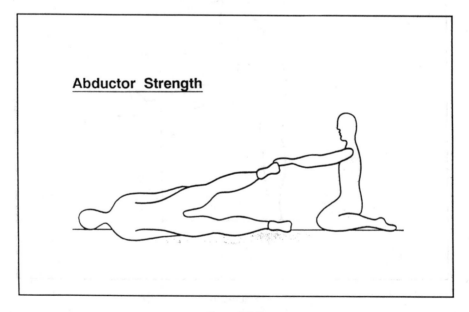

Abductor Strength

Figure 20-6.

Corrective exercise. People who are not able to hold up their legs against their partner's push should exercise daily by laying on their side and raising and lowering the upper leg. Weights can be added to the ankle.

The Iliopsoas Stretch

The Iliopsoas Stretch, see Figure 20-7, should always be included in the workout because there is no criterion test for this exercise. It is one of the most important stretching exercises for improved back protection.

Instructions. Step forward and let the heel of the back foot come off of the ground. Now lower the body weight straight down toward the floor. A stretch is felt in the groin of the back leg. *Do not bounce.* Gently lower the body to

Figure 20-7.

the point that there is pain in the groin, and hold that position while consciously relaxing the pelvis. Both sides must be stretched.

Criterion. There is no criterion for this stretch. This stretch should be included in the daily workout.

Warning. As with all stretching exercises, it is very important not to bounce. Instead, be gentle and relax the area being stretched.

In general, stretch and flexibility are important to everyone. Some people are so tight that even when they do lift properly they do not get much benefit from it because they are too tight or insufficiently strong in one or more of the required muscle groups. They may already have sufficient abdominal strength, but they need to work on flexibility and stretch.

Cardiovascular conditioning.   Cardiovascular conditioning is focused on getting the heart into proper shape so that it can provide adequate nourishment and oxygen to the muscles. Good cardiovascular conditioning requires keeping the pulse rate within the proper range (see below) for 20 minutes at a time, three to four times a week. This can be achieved by swimming, walking, riding a bicycle, or any low-impact exercise. Before beginning this exercise schedule it is important to obtain medical clearance. The proper pulse range for each person is calculated as follows.

- The upper pulse limit is equal to the number 190 minus the person's age.
- The lower pulse limit equals the person's upper limit minus 25.

For example, the upper pulse limit for a 40-year-old person is 150 beats per minute (190 minus 40). The lower pulse limit is 125 beats per minute (150 minus 25). What these numbers mean is that in order for 40-year-old people to improve their cardiovascular condition, three or four times a week for 20–30 minutes at a time they need to exercise at a rate that keeps their pulse between 125 and 150 beats per minute.

Body weight.   Obesity is a problem for someone who must exert himself. For every extra pound of fat, the circulatory system develops an extra mile of capillaries, increasing the load on the heart. In addition, obesity in the form of a pot belly weakens the abdominal muscles, making them ineffective in opposing and strengthening the back muscles. A pot belly also increases the distance that the load is carried from the body, thereby increasing the pressure on the lower back disks. This effect in turn exaggerates the normal curves of the spine. Obesity tends to correlate with a sedentary lifestyle, which also decreases cardiovascular and specific muscle conditioning. Finally, obesity in men increases the amount of body weight that is carried about the waist, adding to the load when they lift or bend. A proper weight management program does not rely on drugs or on fad diets but should include behavioral management principles and the control of fat intake. Such a program can be suggested by a physician for persons who obtain clearance for an exercise program.

Using the Behavioral Observation and Feedback System to Target Back Injuries

ABC Analysis. ABC Analysis can help to answer the question of why people continue to lift improperly even when they "know better." The answer of course is that many natural consequences favor improper lifting. (See Chapters 2 and 16 for other examples of ABC Analysis.) Step 1, consider the common antecedents that workers are aware of: fatigue, weak stomach muscles, weak leg muscles, tight leg muscles, hurrying, ignorance of body mechanics (see Figure 20-8). The natural consequence of improper lifting that workers are often aware of are:

- It is easier on the leg and stomach muscles—soon-certain-positive.
- It conserves energy—soon-certain-positive.
- It saves time—soon-certain-positive.
- Usually there are no noticeable bad effects—soon-certain-positive.
- It feels "natural"—soon-certain-positive.
- There is the possibility of back injury—late-uncertain-negative.
- There is the possibility of being reprimanded—late-uncertain-negative.

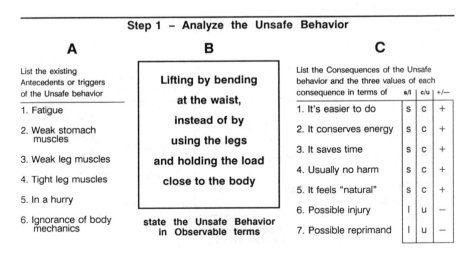

Figure 20-8. Step 1—ABC Analysis of improper lifting.

Of this list of naturally occurring consequences only the last two favor proper lifting. However, notice that these two consequences are of the weakest kind—late, uncertain, and negative. The other consequences in the list are of the strongest kind—soon, certain, and positive—and they all favor unsafe, improper lifting.

The behavior-based approach to back-injury prevention knows better than to rely on the naturally occurring antecedents and consequences of lifting work. Instead, Step 2, it uses training to change the antecedents which trigger worker behavior, Figure 20-9. This training conveys information about body mechanics and introduces physical exercises to increase body strength and flexibility. To bring a change in the consequences, behavioral Observation and feedback are used. Where there is already a behavioral Observation and feedback system in place, Observations are focused on back-related behaviors. Where the behavioral system is not yet in place, it is possible to begin the Observation and feedback process with back-related critical behaviors. The steps are the same as demonstrated in the body of this book: identify → measure → feedback.

Identify. Identify critical back-related behaviors such as:

- Proper body use
- Plan the task
- Get help
- Acknowledge and respond to personal limits
- Warm up

Step 2 – Analyze the Safe Behavior

A	**B**	**C**
List **new** Antecedents – ones that can trigger the Safe behavior	**Lifting by using**	List **new** Consequences – ones that will support the Safe behavior, consequences that are **Soon, Certain, & Positive.**
1. Training in the principles of back-injury prevention	**the legs,**	1. Observation and Self-Observation
2. Criteria testing for employees	**holding the load**	2. Positive feedback from Observers for safe lifting
3. Exercise program for stronger, more flexible physical condition	**close to the body**	3. Positive feedback from supervisors for safe lifting 4. Improved % Safe ratings for the workforce

state the Safe Behavior
in Observable terms

Figure 20-9. Step 2—analyze proper lifting.

Back–Injury Prevention
Observation Data Sheet

1.0 Before You Act	Safe	Unsafe
1.1 Warm–Ups		
1.2 Plan	_____	_____
1.3 Get Help		

2.0 When You Act		
2.1 Test First	_____	_____
2.2 Maintain Curves	_____	_____
2.3 Minimize Impact	_____	_____
2.4 Rest from Repetitions	_____	_____
2.5 Load Close	_____	_____
2.6 Range of Activity	_____	_____
2.7 Footing / Path	_____	_____
2.8 Turn vs. Twist	_____	_____

3.0 At Other Times		
3.1 Exercise		
3.2 Stretch	_____	_____
3.3 Meet Criteria		

Figure 20-10. Back-Injury Prevention—Observation data sheet.

- Stretch
- Exercise
- Rest from repetitive motion
- Self-Observation.

Write operational definitions as well as an Observation Data Sheet. Prepare a Data Sheet (see Figure 20-10).

Measure. Train Observers to Observe for back-related behaviors. Slides should be used to show the behaviors which have been operationally defined. Observers are trained and checked for reliability — correspondence of %Safe between two Observers who are Observing the same behavior. Employees at high risk for back injury should be trained in Self-Observation of their critical behaviors, see Chapter 18.

Feedback. Verbal feedback is provided on the spot by the Observer, emphasizing the positive, especially areas of improvement. Safe behavior percentages are charted to show progress (or problems) over time. Data is analyzed over time to provide additional feedback, to identify new problems, and to provide a basis for problem-solving.

SUMMARY

Once again as in the special application of managing employees who have had multiple accidents, back-injury prevention represents the whole behavior-based process focused on a specialized area that has the potential to yield very high benefits.

Bibliography

The articles and books in this bibliography discuss the use of behavioral science methods for accident prevention. The authors provide this list for readers who want more in-depth information on the subject.

Chhokar, S. J., and J. A. Wallin. 1984. "Improving safety through applied behavior analysis." *Journal of Safety Research* 15(4): 141-51.

Cohen, H. H., and R. C. Jensen. 1984. "Measuring the effectiveness of an industrial lift truck safety training program." *Journal of Safety Research* 15(3): 125-35.

Fellner, D. J., and B. Sulzer-Azaroff. 1984. "Increasing industrial safety practices and conditions through posted feedback." *Journal of Safety Research* 15(1): 7-21.

Geller, E. S. 1984. "A delayed reward strategy for large-scale motivation of seat belt use: a test of long-term impact." *Accident Analysis and Prevention* 16(5/6): 457-63.

Haynes, R. S., R. G. Pine, and H. G. Fitch. 1982. "Reducing accident rates with organizational behavior modification." *Academy of Management Journal* 25(2): 407-16.

Hidley, J. H., and T. R. Krause. December, 1988. "Exceptional safety programs can be dangerous to your health." *Occupational Health and Safety.*

Hoyos, C. G., and B. Zimolong. 1988. *Occupational Safety and Accident Prevention, Behavioral Strategies and Methods.* NY: Elsevier.

Kim, J., and C. Hamner. 1982. "Effect of performance feedback and goal setting on productivity and satisfaction in an organizational setting." *Journal of Applied Psychology* 61: 48-57.

Komaki, J., K. D. Barwick, and L. R. Scott. 1978. "A behavioral approach to occupational safety: pinpointing and reinforcing safety performance in a food manufacturing plant." *Journal of Applied Psychology* 63(4): 434-45.

Komaki, J., A. T. Heinzmann, and L. Lawson. 1980. "Effect of training and feedback: component analysis of a behavioral safety program." *Journal of Applied Psychology* 65(3): 261-70.

Komaki, J., R. L. Collins, and P. Penn. 1982. "The role of performance antecedents and consequences in work motivation." *Journal of Applied Psychology* 67(3): 334-40.

Krause, T. R., J. H. Hidley, and W. Lareau. July, 1984. "Behavioral science applied to industrial accident prevention." *Professional Safety.*

Krause, T. R., and J. H. Hidley. 1990 (in press). "A behavior-based safety management process." *Applying Psychology in Business: The Manager's Handbook.* Lexington Books.

Krause, T. R., J.H. Hidley, and D. Schorr. February, 1988. "Managing safety means focusing on behavior." *PIMA Magazine* 70(2): 45-48.

Krause, T. R., J. H. Hidley, and S. J. Hodson. December, 1988. "Behavioral science in the workplace: techniques for achieving an injury-free environment." *Modern Job Safety and Health Guidelines.* Prentice Hall Information Services.

Krause, T. R., and J. H. Hidley. August, 1989. "Improving safety culture—discipline isn't enough." *Insights Into Management.* National Safety Management Society.

Krause, T. R., and J. H. Hidley. October, 1989. "Behaviorally based safety management: parallels with the quality improvement process." *Professional Safety* 20-25.

Krause, T. R., and J. H. Hidley. October, 1989. "Improving safety culture—behavior versus attitude." *Insights Into Management.* National Safety Management Society.

Krause, T. R., and J. H. Hidley. December, 1989. "Improving safety culture—the critical-mass approach." *Insights Into Management.* National Safety Management Society.

Levensen, H., et al. 1979. "Industrial accidents and recent life events." *Journal of Occupational Medicine* 21: 26-31.

Levi, L. 1979. "Occupational mental health: its monitoring, protection, and promotion." *Journal of Occupational Medicine* 21: 26-32.

Luthans, F., and R. Kreitner. 1975. *Organizational Behavior Modification.* Scott Foresman.

Ouchi, W. G. 1981. *Theory Z: how American business can meet the Japanese challenge.* MA: Addison-Wesley.

Petersen, D. 1981. *Human Error Reduction and Safety Management.* Garland Press.

Petersen, D. 1984. "An experiment in positive reinforcement." *Professional Safety* 29(4): 30-35.

Petersen, D. 1989. *Safe Behavior Reinforcement.* Aloray, Inc.

Reber, R. A., and J. A. Wallin. 1983. "Validation of behavioural measure of occupational safety. "*Journal of Organizational Behaviour Management* 5(2): 69-77.

Reber, R. A., J. A. Wallin, and J. S. Chhokar. 1984. "Reducing industrial accidents: a behavioural experiment." *Industrial Relations* 23(1): 119-25.

Rhoton, W. W. 1980. "A procedure to improve compliance with coal mine safety regulations." *Journal of Organizational Behaviour Management* 2(4): 243-49.

Schaeffer, M. 1976. "An evaluation of epidemiologic studies related to accident prevention." *Journal of Safety Research* 8: 19-22.

Smith, M. J., W. K. Anger, and S. S. Uslan. 1978. "Behavioral modification applied to occupational safety." *Journal of Safety Research* 10(2): 87-88.

Sulzer-Azaroff, B., and M. C. de Santa Maria. 1980. "Industrial safety reduction through performance feedback." *Journal of Applied Behavior Analysis* 13: 287-95.

Sulzer-Azaroff, B., and D. J. Fellner. 1984. "Searching for performance targets in the behavioral analysis of occupational health and safety: an assessment strategy." *Journal of Organizational Behaviour Management* 6(2): 53-65.

Zohar, D. 1980. "Promoting the use of personal protective equipment by behavior modification techniques." *Journal of Safety Research* 12(2): 78-85.

Zohar, D., and N. Nussfeld. 1981. "Modifying earplug wearing by behaviour modification techniques: an empirical evaluation." *Journal of Organizational Behaviour Management* 3(2): 41-52.

Index

Index